Strategies of Knowledge Transfer for Economic Diversification in the Arab States of the Gulf

Edited by Rasmus Gjedssø Bertelsen, Neema Noori and Jean-Marc Rickli

Gerlach Press

First published 2018
by Gerlach Press
Berlin, Germany
www.gerlach-press.de

Cover Design: Frauke Schön, Hamburg
Set by Anne Jeschke, Gerlach Press
Printed and bound in Germany

British Library Cataloguing in Publication Data.
A catalogue record for this book is available from the British Library.

Bibliographic data available from Deutsche Nationalbibliothek
http://d-nb.info/1129886212

ISBN: 978-3-95994-014-6 (hardcover)
ISBN: 978-3-95994-015-3 (eBook)

Strategies of Knowledge Transfer for Economic Diversification in the Arab States of the Gulf

The Gulf Research Centre Book Series at Gerlach Press

The Gulf Region: Economic Development and Diversification (4 vols set)
Ed. by L. Luciani, S. Hertog, E. Woertz and R. Youngs
ISBN 9783940924001, 2012

The GCC in the Global Economy
Ed. by Richard Youngs
ISBN 9783940924018, 2012

Resources Blessed: Diversification and the Gulf Development Model
Ed. by Giacomo Luciani
ISBN 9783940924025, 2012

GCC Financial Markets: The World's New Money Centers
Ed. by Eckart Woertz
ISBN 9783940924032, 2012

National Employment, Migration and Education in the GCC
Ed. by Steffen Hertog
ISBN 9783940924049, 2012

Asia-Gulf Economic Relations in the 21st Century: The Local to Global Transformation
Ed. by Tim Niblock with Monica Malik
ISBN 9783940924100, 2013

A New Gulf Security Architecture: Prospects and Challenges for an Asian Role
Ed. by Ranjit Gupta, Abubaker Bagader, Talmiz Ahmad and N. Janardhan
ISBN 9783940924360, 2014

Gulf Charities and Islamic Philanthropy in the 'Age of Terror' and Beyond
Ed. by Robert Lacey and Jonathan Benthall
ISBN 9783940924322, 2014

State-Society Relations in the Arab Gulf States
Ed. by Mazhar Al-Zoby and Birol Baskan
ISBN 9783940924384, 2014

Political Economy of Energy Reform: The Clean Energy-Fossil Fuel Balance in the Gulf
Ed. by Giacomo Luciani and Rabia Ferroukhi
ISBN 9783940924407, 2014

The Silent Revolution: The Arab Spring and the Gulf States
Ed. by May Seikaly and Khawla Mattar
ISBN 9783940924346, 2014

Security Dynamics of East Asia in the Gulf Region
Ed. by Tim Niblock with Yang Guang
ISBN 9783940924483, 2014

Islamic Finance: Political Economy, Performance and Risk
Ed. by Mehmet Asutay and Abdullah Turkistani
ISBN 9783940924124, 3 vols set, 2015

Employment and Career Motivation in the Arab Gulf States: The Rentier Mentality Revisited
Ed. by Annika Kropf and Mohamed Ramady
ISBN 9783940924605, 2015

The Changing Energy Landscape in the Gulf: Strategic Implications
Ed. by Gawdat Bahgat
ISBN 9783940924643, 2015

Sustainable Development Challenges in the Arab States of the Gulf
Ed. by David Bryde, Yusra Mouzughi and Turki Al Rasheed
ISBN 9783940924629, 2015

The United States and the Gulf: Shifting Pressures, Strategies and Alignments
Ed. by Steven W. Hook and Tim Niblock
ISBN 9783940924667, 2015

Africa and the Gulf Region: Blurred Boundaries and Shifting Ties
Ed. by Rogaia Mustafa Abusharaf and Dale F. Eickelman
ISBN 9783940924704, 2015

Rebuilding Yemen: Political, Economic and Social Challenges
Ed. by Noel Brehony and Saud Al-Sarhan
ISBN 9783940924681, 2015

Science and Technology Development in the Gulf States: Economic Diversification through Regional Collaboration
Ed. by Afreen Siddiqi and Laura Diaz Anadon
ISBN 9783959940023, 2016

Higher Education Investment in the Arab States of the Gulf: Strategies for Excellence and Diversity
Ed. by Dale F. Eickelman and Rogaia Mustafa Abusharaf
ISBN 9783959940122, 2016

Iran's Relations with the Arab States of the Gulf: Common Interests over Historic Rivalry
Ed. by Maaike Warnaar, Luciano Zaccara and Paul Aarts
ISBN 9783959940047, 2016

Egypt and the Gulf: A Renewed Regional Policy Alliance
Ed. by Robert Mason
ISBN 9783959940061, 2016

Intellectual Property Rights: Development and Enforcement in the Arab States of the Gulf
Ed. by David Price and Alhanoof AlDebasi
ISBN 9783959940108, 2016

The Arab States of the Gulf and BRICS: New Strategic Partnerships in Geopolitcs and Economics
Ed. by Tim Niblock, Alejandra Galindo and Degang Sun
ISBN 9783959940085, 2016

The Arms Trade, Military Services and the Security Market in the Gulf States: Trends and Implications
Ed. by David B. Des Roches and Dania Thafer
ISBN 9783959940160, 2016

Strategies of Knowledge Transfer for Economic Diversification in the Arab States of the Gulf
Ed. by Rasmus Gjedssø Bertelsen, Neema Noori and Jean-Marc Rickli
ISBN 9783959940146, 2018

Yemen and the Gulf States: The Making of a Crisis
Ed. by Helen Lackner and Daniel Martin Varisco
ISBN 9783959940306, 2018

Conflict Resolution and Creation of a Security Community in the Gulf Region
Ed. by Tim Niblock with Talmiz Ahmad and Degang Sun
ISBN 9783959940368, 2018

The Future of Labour Market Reform in the Gulf States: Towards a Multi-Disciplinary, Evidence-Based and Practical Understanding
Ed. by David B. Jones and Sofiane Sahraoui
ISBN 9783959940344, 2018

Gulf Research Centre Cambridge
Knowledge for All

Contents

Preface

The idea for this volume emerged from a series of transnational collaborations that began in 2010 through a chance meeting of Rasmus Gjedssø Bertelsen and Neema Noori at an International Studies Association Meeting in New Orleans, Louisiana. At the time, Rasmus was transitioning from a post-doctorate at Harvard to a postdoc at United Nations University-Institute of Advanced Studies in Yokohama studying early 19th century intellectual exchange networks linking the United States and Western Europe with the Middle East, Japan, and China. Neema, on the other hand, had just begun working and writing about the proliferation of American-style universities and branch campuses in the Gulf. Though Rasmus' research focus was more historically oriented, Neema and Rasmus shared a common interest in transnational higher education networks and the Middle East. With his background in small state security studies and his previous positions at the Joaan Bin Jassim Joint Command and Staff College in Doha and at Khalifa University in Abu Dhabi, fellow editor Jean-Marc Rickli played a pivotal role in encouraging us to think beyond conceptualizing universities as the central nodes in transnational knowledge networks, particularly as they pertained to the Gulf. This book, therefore, is an attempt to widen our focus beyond traditional academic institutions, and, perhaps more importantly, beyond the West as we have also solicited contributions that explore intellectual relationships that connect the Gulf to the East.

Many people have contributed to this book, both directly and indirectly, and we would like to take this opportunity to thank them. Certainly, none of this would have been possible were it not for the generous support of the Gulf Research Centre. Our gratitude also extends to our home universities, the University of West Georgia and the University of Tromsø-The Arctic University of Norway, for providing the support that enabled us to meet in the first place. We also owe a debt of gratitude to Lacey Moore, a graduate student at the University of West Georgia, who generously invested a great deal of time and energy in helping us edit and format the book chapters. Angela Turner, Amber Locke, and Jordan Smith also helped edit select chapters. Finally, we are very grateful to the participants for working diligently to refine their contributions for publication.

Please note that this was a collaborative effort. Each of the editors contributed equally to the project and the names of the editors are ordered alphabetically.

For their enduring support, this book is dedicated to María Fernanda González Rojas, Susan Hackett-Noori and Jean Rickli.

Rasmus Gjedssø Bertelsen, Neema Noori and Jean-Marc Rickli

1

Transnational Knowledge Relations for Building Knowledge-Based Societies in the Gulf

Rasmus Gjedssø Bertelsen, Neema Noori and Jean-Marc Rickli

The precipitous decline in the price of oil over the past two years has served as a cautionary tale for the Gulf Cooperation Council states. Low oil prices are causing severe economic and fiscal problems for these states and threatening a socio-political order built on redistributing oil and gas rents for political allegiance. The region's political and economic importance also make these fiscal challenges and socio-political developments a concern for countries around the world. Regional governments are aware of the immediate fiscal challenges of low oil prices and the long-term challenges of diversifying oil- and gas-based national economies. The ruling elite have for years considered developing knowledge-based economies as a means to diversify and create sustainable post-fossil-fuel-based economies. Transnational knowledge relations and talent mobility have been and remain key instruments for pursuing knowledge-based Gulf societies. This book analyzes and discusses the experiences of transnational knowledge workers in a variety of arenas including higher education, healthcare, business, and government. In pursuing this focus, this book's contributors address two questions that are central to the region's goals of developing a knowledge-based economy. First, is it possible to build a knowledge-based economy by relying almost exclusively on importing capacity? Second, what patterns of transnational knowledge relations are most likely to produce successful outcomes?

The forces responsible for this steep drop in oil prices, namely sluggish growth in China and other emerging economies as well as a deliberate new policy of OPEC spearheaded by Saudi Arabia to defend market share instead of price stability as a means to counter the competition of US shale oil and gas producers[1], have forced many GCC economies to consider policy tools that were once unthinkable and that call into question the structural nature of the rentier state. For example, Saudi Arabia has begun cutting subsidies for gas, water, and electricity and the GCC as a whole has begun to debate the adoption of a regional value added tax (Diaa, 2016; al-Hatlani, 2016). The subject of this book, which concerns the role of transnational knowledge relations in developing knowledge-based societies and economies in the Gulf, is a timely one. For the past two decades, Gulf states such as the U.A.E. and Qatar have made significant progress towards diversifying their economies: oil and natural gas make up less than 70% of government revenues in these two countries (Rickli 2014; Charles 2013a; Young 2016). This effort is clear from the great strides both nations have taken in recent decades to strengthen their capacities in education and research.

Organizational Logics of Knowledge-Based Economies

However, the purposeful construction of a knowledge economy requires more than the diversification of economic activities. It requires a wholesale adaptation of social, political, and economic institutions to support an economy based on intellectual capital rather than natural resources or physical capital. As such, this transition marks a paradigmatic shift from the tangible to the intangible in terms of what is conceived of as having value (Powell and Snellman, 2004: 201). Powell and Snellman define the knowledge economy as "production and services based on knowledge-intensive activities that contribute to an accelerated pace of technological and scientific advance as well as an equally rapid obsolescence" (Powell and Snellman, 2004: 201). Economic sociologists have commented on how the organizational environments engendered by the knowledge economy are less likely to be hierarchical and centralized. As opposed to the Fordist industrial workplace, in which workers were subject to high levels of discipline and control, workers "in the new economy" have more autonomy and flexibility in terms of where and how they perform their jobs. As such, organizational social scientists are quick to point out that potential gains from the knowledge economy cannot be realized unless "new technologies are married to complementary organizational practices" (Powell and Snellman 2004: 215). In other words, for the context of this book, importing technology, knowledge workers, and the transfer of new forms of knowledge are not on their own sufficient to enabling a knowledge economy.

Creating a knowledge economy requires the adoption of new organizational cultures. David Stark (2011: 25) coined the term heterarchy as a way of distinguishing these new organizational forms, which are decentralized, non-hierarchical, and often lack clear boundaries to distinguish divisions both within and outside the firm. Because of a perpetually shifting business landscape, organizations that have adapted to the knowledge economy, whether in Silicon Valley or Wall Street, have to remain nimble to take advantage of new opportunities unfolding in an age of persistent uncertainty in which a business organization's fundamental activity may change drastically. For example, GM is best known as a car manufacturing corporation. Increasingly, however, as Stark points out, so much of GM's research and development expenditures are devoted to enhancing navigation and entertainment systems within the automobile, that it could easily be categorized as a computer business. Similarly, Apple's reported flirtation with developing a self-driving car could alter our understanding of its core business identity. Consequently, a core theme of many of the chapters in this volume concerns the ability of Gulf societies and economies to promote an institutional environment that will support the growth of organizational cultures that hold the greatest promise of stimulating innovation (Harris, 2015).

Geopolitical Context

The wider Middle East and the GCC states have played a central role in international politics and economics for decades. The Middle East has been and continues to be the scene of conflicts causing great human suffering, currently in places such as Iraq, Syria and Yemen. There is a renewed geopolitical competition between Iran and the Arabian Peninsula monarchies led by Saudi Arabia with wide-ranging regional and global implications (Rickli, 2016a). Aside from internal competition for regional hegemony, the Middle East has since the Cold War also served as a site for great power rivalry. Despite Russia's military intervention in the Syrian civil war in 2015, the US has for decades remained the region's hegemonic power. This was formalised by the so-called 1980 Carter doctrine which states that any "attempt by outside force to gain control of the Persian Gulf region will be regarded as an assault on the vital interests of the United States" and as such will "be repelled by any means necessary, including military force" (Carter, 1980). Mirroring these military and diplomatic ties with the US, the GCC states have over decades built dense transnational knowledge relations with the West, particularly the United States. Myriads of Western academics study the GCC countries in universities, think tanks and public and private institutions. US civilian and military government departments in particular have deep experience working with Gulf issues. Private companies have decades of work experience in the region, where, for instance, American oil companies have since the interwar periods been deeply engaged in the exploration and production of oil and gas. On the GCC side, many decision-makers in government, business, higher education and other sectors of society have post-secondary degrees from American or other Western institutions.

There are also non-Western sources of transnational knowledge relations in the Gulf, largely absent in the existing literature, that deserve close attention. The Gulf is historically at the cross-roads of East and West, with strong historic trade and cultural relations towards the East over the Indian Ocean and overland to Asia. This key location on the routes connecting East and West helps explain the British Empire's strategic interest in the Gulf. The Gulf is also deeply affected by broader global trends, namely the secular and seemingly ineluctable transfer of power from Western states to Eastern states and the diffusion of power from state to non-state actors. This Eastward power transition is largely driven by China's spectacular economic growth. While seeking to avoid a concurrent political and security involvement, China has become a very large energy customer in the Gulf. From 2003 to 2013, China accounted for 60% of the growth in the global energy demand (Daojiong and Meidan, 2015: 2). China imports about 55% of its oil from the Gulf and is the biggest importer of oil from the GCC (Andersen and Jiang, 2015: 24).

President Xi Jinping's rise to power in 2012 coincided with an aggressive expansion of economic engagement with the region. From investments in the oil and gas industry, major infrastructural projects such as the One Belt, One Road project to its attempts to enter the halal foods industry, China has made the region a geopolitical priority (Allen-Ebrahimian, 2016). The most ambitious of these initiatives is the One Belt, One Road project, a multi-phase effort to build new transportation linkages to connect the disparate parts of the Eurasian landmass to China. Similarly, the increasing Chinese presence along the Asian and African coastlines contributes to the creation of the so-called "String of Pearl" from China to Africa.

Envisioned as a 21st century version of the silk road, China has also invested in building the institutional infrastructure to facilitate cultural and educational exchange along the proverbial silk highway. From Arabic language training and Middle East cultural institutes to World Muslim City in Yinchuan, a kitschy amusement park, Chinese authorities are investing in the multidimensional transnational cultural and educational networks and resources that may contribute to strengthening noncommercial ties between the two regions (Haddad-Fonda, 2016). Similarly, Japan has been a long-time large energy customer in the Gulf. It too has invested in cultural and intellectual relations with the region. One particular strength of this book is that it includes chapters on both Chinese and Japanese transnational knowledge relations with the Gulf.

The GCC states have for decades been key participants in Middle East geopolitics and geoeconomics. The region's mineral resource wealth has afforded relatively small countries such as the United Arab Emirates and Qatar the capability to project power far beyond what one would expect of countries with their geographic size and population numbers (Al-Mezaini and Rickli, 2016; Rickli, 2014; Rickli, 2016b). Whether due to the astute decision-making of the ruling elite, low-wage immigrant labor, or the security guarantees provided by US military power, Gulf city states such as Qatar, which is slated to hold the FIFA World Cup in 2022 and has financed the creation of Al Jazeera – an international news agency that rivals CNN and BBC in terms of global popularity and coverage – have acquired vast influence disproportionate to their size (Kamrava, 2015). Mehran Kamrava aptly refers to Qatar as a "tiny giant," while U.S. generals refer to the United Arab Emirates (UAE) as "Little Sparta" because of its proven ablitity to project military power in the region (Chandrasekaran, 2014). The UAE also hosts two of the world's largest international airline companies, Emirates and Etihad Airways, as well as the third largest port operations company, DP World. Both Qatar and the UAE also have among the largest sovereign wealth funds in the world (Rickli and Charles, 2013b). Importantly, for the purposes of this book, their achievements have not gone unnoticed as countries such as Kazakhstan, Iran, and Azerbaijan, to name a few, have tried to replicate the grandiose and spectacular urban projects that were pioneered in the region. Therefore, countries such as Qatar and the

United Arab Emirates have, in their own right, become important nodes in trans-regional knowledge dissemination networks.

The socio-political-economic development of these countries is therefore a topic of key importance for their own populations, the wider Middle East and the world. Transnational knowledge relations between the GCC states and the outside world are important factors both for domestic development and for their role in regional and global political and economic affairs. Rather than focus on diplomatic or military transnational knowledge relations, this book's focus is on how those relations contribute to economic development (See Al-Mezaini and Rickli, 2016). As such, it represents an important contribution to the study of these transnational knowledge relations.

Nationalization Programs and Transnational Knowledge Networks

Economic diversification is the overriding policy goal of the GCC countries, where building knowledge-based economies using transnational knowledge relations, talent mobility, and the empowerment of the local population through nationalization programs (e.g. Qatarisation or Emiratisation) are seen as a key policy instruments. Traditionally, rather than relying exclusively on building indigenous competencies to support efforts to diversify their economies; most countries in the GCC have depended heavily on importing capacity, usually in the form of expatriate technical and managerial talent. Resource rich and population poor, GCC states have historically recruited teachers, doctors, engineers, and even clerics from other countries. In doing so they are continuing a pattern that was established under British colonialism. This is a comparatively novel approach; most successful emerging economies over the past five decades such as Korea, Taiwan, and Chile have pursued a more inward focused development strategy opting instead to recruit and train members of their own population to staff universities, hospitals, and businesses. By contrast, the Gulf's recipe for economic development has rested on a combination of abundant oil resources, cheap labor, and external security guarantees. Yet, as the nationalization programs referred to previously demonstrate, this pattern has begun to change.

The focus of this book is on the agents that serve as conduits for the regional dissemination of expertise, both technical and otherwise. In pursuing this question, the contributors to the book address the two questions defined at the outset which are central to the region's goals of developing a knowledge-based economy. First, is it possible to build a knowledge-based economy by relying almost exclusively on importing capacity? And, if not, what are the challenges to developing the skills and capabilities of the local population? Second, what patterns of transnational knowledge relations are most likely to produce successful outcomes? In tackling these two questions, the contributors to this volume primarily focus on the interface between the local and the global. In treating these intersections, as the focus of our analysis, our contributors examine both the synergies and the inevitable friction that result from attempts to forge cross-national and cross-cultural intellectual relationships. And, importantly, they examine the ways in which the resulting connections influence patterns of knowledge diffusion that vary across time and space.

Research on universities as transnational actors demonstrates that these institutions hold the promise of developing relations with a wide range of actors across borders, such as academics, business elites, civil society actors, philanthropic organizations, and political elites. The private, transnational universities in the Gulf and the wider Middle East are to a varying extent "information and resource bridges" that facilitate heavy traffic of knowledge, talent, and financial resources in both directions (Bertelsen, 2014b; Bertelsen et al., 2014; Bertelsen et al., 2013; Bertelsen et al., 2013; Bertelsen, 2012b; Bertelsen, 2012a; Bertelsen, 2009b; Bertelsen, 2009a; Bertelsen et al., 2014; Noori, 2014; Noori and Anderson, 2013).

Importantly, the contributors to this book go beyond educational institutions in their analyses. In addition to universities, the authors of this volume explore the role of transnational knowledge relations through think tanks, international experts, and corporations with an emphasis on promoting the mobility of human capital between the Gulf and the West for developing knowledge-based societies and economies in the region. As such, the central focus of the book is on the transnational networks for intellectual exchange and talent mobility in the Gulf and of the importance of transnational knowledge exchange networks for socio-economic development. The authors of this book define transnational knowledge relations as a set of transnational networks mobilizing knowledge workers, academic institutions, business organizations, and concpetual frameworks. To date, research on transnational knowledge relations has focused heavily on relations between Western and Gulf-based universities. This attention has partly been driven by a political-strategic concern with soft power and public diplomacy during the War on Terror (Djerejian, 2003; Lord, 2006; Rice, 2006; Rugh, 2006; Armitage and Nye, 2007; Council on Foreign Relations, 2007; Peled, 2012).

The other main motivator of academic attention to transnational knowledge relations with the Gulf has been on socio-economic development in the wider Middle East, where the United Nations Development Programme Arab Human Development report in 2003 highlighted the critical state of human capital in the Arab world and how it impeded human development (United Nations Development Programme, 2003). Compared to other Arab states, the GCC states are in a privileged fiscal position because of oil and gas revenues. However, as pointed out throughout this book, the GCC states are hard pressed to diversify their economies from oil and gas rents: a decades long policy focus that has met with limited success (Al-Atiqi et al., 2009; Abi-Mershed, 2010; Al-Hamarneh, 2006; Al-Lamki, 2006; Beck, 2006; Davidson and Smith, 2008; Herbert Quandt Stiftung 2006; Lane, 2010; Lewin, 2009; Mazawi and Sultana, 2010; Ministry of Higher Education. Sultanate of Oman , Office Méditerranéen de la Jeunesse; Olson, 2009; Olson, 2008b; Olson, 2008a, The Middle East Institute, 2010; Thomson, 2004).

This book will contribute to the above mentioned scholarship on Gulf and Middle East education, research development, and innovation within a global literature on the role of mobility, brain circulation, human capital and knowledge for innovation, entrepreneurship and socio-economic development, which is, for instance, the focus of World Bank Group work on Education for the Knowledge Economy generally and for the Arab world specifically (OECD, 2008; OECD, 2002; Saxenian, 2006; Saxenian, 2005; Saxenian and Sabel, 2008; Seguin et al., nd; Zweig et al., 2004; Kuznetsov, 2006; The World Bank Group, 2009; Altbach and Salmi, 2011; World Bank, 2008). Higher education and research is global in nature and has a strong transnational element, which is clearly reflected in current efforts to develop tertiary education in the Gulf as a whole. Many of the new higher education and research institutions created in Gulf societies are based on transnational pacts with higher education institutions in North America, Western Europe, or Australia for purposes of quickly assuring quality in education and research (Bertelsen and Noori, nd; Bertelsen, 2012a; Bertelsen, 2009b; Noori and Anderson, 2013; Noori and Anderson, 2011). Education City in Qatar stands out as one clear example of a transnational strategy for rapidly building quality higher education and research in a Gulf society.

Higher education and research institutions in the Gulf contribute significantly to the development of knowledge-based societies through building strong transnational relations with a range of private, public and civil society actors. Higher education and research institutions have historically played key roles as what Bertelsen has coined "information and resource bridges" between societies. As such, the recent proliferation of joint ventures and the influx of Western institutions in the region is not altogether a new phenomenon. The Middle East has a strong history of transnational higher education institutions stretching back to, e.g., Robert College in Istanbul (1863), Syrian Protestant College (1866)/American University of Beirut (1920), Lebanese American University (1924), Université Saint Joseph de Beyrouth (1875) and American University in Cairo (1919). Since the mid-1990s, the Gulf has been at the forefront of efforts to expand transnational educational linkages with Western institutions such as La Sorbonne and New York University in the UAE or Georgetown and Carnegie Mellon Universities in Qatar.

Both sets of cases, the contemporary and the historic, demonstrate how universities can interface with a range of private, public and civil society actors from both the Middle East and the West. These universities facilitate the circulation of leading academics in the West enabling the transfer of information, ideas and talent back and forth between regions. They connect with leading international companies contributing to the development of the business sector in the Middle East and creating job opportunities for Middle Eastern youth. They link with overseas political and governmental networks, which in turn inform policy debates. These universities have long standing partnerships with leading civil society organizations and large philanthropic foundations (Bertelsen and Noori, nd; Bertelsen, 2014b; Bertelsen, 2012b; Bertelsen, 2012a; Bertelsen, 2009b;

Bertelsen, 2009a; Noori, 2014; Noori and Anderson, 2013; Noori and Anderson, 2011; Bertelsen, Forthcoming; Bertelsen, 2014a).

A key question concerns the degree to which the long-term reliance on mineral resource exports exerts path-dependent effects on future economic reform, impeding the region's goals of developing a knowledge-based economy. Much of the debate on the rentier state has focused on the various ways in which mineral resource dependence tends to suppress political reform. Research in this vein has also shown that mineral resource dependence has historically undermined attempts at economic diversification. Drawing on Beblawi's work on the so-called "rentier mentality," Hvidt argues that the generous welfare states, themselves a product of the rentier economy, do not effectively create an adequate disciplinary environment for Gulf citizens (Hvidt, 2015). In other words, the hard work required to pursue an elite education is not adequately incentivized. Since education forms the bedrock of knowledge-based economies and societies, this failure may comprise the ability of these states to successfully embark on this trajectory.

Furthermore, nationalization programs with the stated aim of reducing dependence on expatriate workers are at times in conflict with initiatives to internationalize higher education, such as Emiratization in the UAE and Qatarization in Qatar. To illustrate this point, very few of the full time faculty at universities such as Zayed University in Dubai, the American University of Sharjah, or the American University of Kuwait are Gulf citizens. Therefore, existing labor laws designed to both facilitate the importation of foreign workers while maintaining control of the terms of their stay, create suboptimal outcomes for two principle reasons. First, imported academics very often stay for a short period of time (3-5 years), which prevents the establishment of institutional memory. Second, by relying mostly on imported academics, the development of national experts is undermined. This structural problem is not limited to academia but is consubstantial to the working environment in the Gulf.

Purpose and Structure

The chapters in this volume draw from a wide range of disciplines and vary in terms of their theoretical assumptions and the methodological toolkits they deploy. With scholars contributing both from within and without the region, the book offers perspectives that are sensitive to local context, but that also draw perspectives from non-gulf locales. Importantly, our contributors include both academic and non-academic experiences. Their work address the role of transnational knowledge relations in meeting the fundamental economic policy challenge of the GCC: the transition from oil- and gas-based economies to more knowledge-based economies. The chapters deal with the two questions of, first, is it possible to build a knowledge-based economy by relying almost exclusively on importing capacity? Second, what patterns of transnational knowledge relations are most likely to produce successful outcomes? The topic and questions of the book are addressed at three levels:

a public policy level, a company level, and an individual level. Thereby the book touches upon three important decision-making levels for this transition, while supplementing the previous attention to the higher education sector.

We have organized the book into three parts. The first part is oriented primarily towards the nation-state level and looks at how public policy can be used to promote a knowledge-based economy. The second part is focused on national and transnational business organizations. The third and final part examines how macro-level institutional and political forces resonate at the personal level.

At the public policy level, the authors, importantly, address a wider set of issues beyond the role of higher education policy for stimulating knowledge-based economic development. Ali A. Alraouf discusses the role of the city of Doha, urbanity, and urban development for building a Qatari knowledge-based economy. By drawing from leading theorists of the metropolis under globalization such as Saskia Saseen and Manuel Castells, Alraouf makes a compelling case for the centrality of cities in the GCC in engendering a knowledge-based economy. Urban centers are the nodes of a globalized knowledge-based economy where information, talent and resources flow between centers of finance, specialized business services, higher education, research, development, and innovation. Importantly, Alraouf's emphasis on how innovative urban design principles can serve as both a vehicle for attracting intellectual talent and a means to foster collaborative exchange within the city, a longstanding hallmark of the most dynamic global cities. A critical component of these efforts has to do with investments in cultural institutions such as museums to serve as public amenities for the growing number of knowledge workers that Qatar hopes to attract.

By exploring large-scale investment in the infrastructure of knowledge-generative institutions, Pia Anderson's chapter addresses a related theme. However, by expanding the scope of her analysis beyond higher education policy, Anderson's chapter addresses a gap in the existing literature on regional efforts to build a knowledge-based economy. Anderson's chapter covers lesser known but equally impactful projects such as book fairs, archival centers, museums, and science centers. In analyzing the functional objectives of these projects, Anderson also addresses the symbolic goals of these initiatives and their capacity to inspire, marshall support from domestic audiences, attract global attention, and serve as trophies in competitive relations with neighboring city-states.

Daniel Goldman's chapter provides a comprehensive assessment of the region's attempts to diversify away from its dependence on fossil fuel exports. Drawing from wide-ranging case studies of countries that have made significant progress in developing high technology export economies such as South Korea, China, India, and Rwanda, Goldman highlights important policy tools that could be deployed to stimulate a knowledge-based economy in the region. While focused mainly on how to deepen the regional talent pool and what technologies to invest in, Goldman provides a set of concrete policy recommendations to promote sustainability and advanced technology in the region.

Eric Baark's chapter explores the role that universities can play in fostering the capacity for innovation in Gulf states. Baark's chapter adopts a comparative approach in examining the kinds of productive relationships, both national and international, that can support national innovation systems. In particular, Baark emphasizes the need for regional universities to broaden their domestic outreach and serve as vehicles for the regional dissemination of knowledge. Baark also formulates a set of policy proposals for this purpose.

Isam Hamza shows how Japan has, since the global oil crisis of the 1970s, used higher education relations with the Middle East for foreign and energy-security reasons. Japan is a major energy customer in the GCC, and Japan's energy security is heavily dependent on its political relations with the GCC. Japan first worked through Cairo University to develop a Japanese language program with the goal of producing a cadre of Arab intermediaries with Japanese linguistic expertise to broaden intellectual, economic, and political relations between the two countries. Graduates from the Cairo University program were later used to develop Japanese studies programs in other GCC countries. Drawing from personal experience as one of the early participants in these transnational programs, Hamza's chapter provides an insightful analysis of the centrality of higher education relations to Japanse diplomatic strategies in the region.

Camilla Nellemann, Ibrahim Radhi, and Gao Xudong each focus on how business organizations can serve as vehicles for the transmission of knowledge and technical expertise. Camilla Nelleman's chapter explores how the organizational configuration of Japanese business conglomerates influences their collaborative ventures in the GCC. Drawing from two case studies, one involving a Danish multinational company and its partnership with Mitsubishi and the other involving SBI and its partnership with a business conglomerate in Bahrain, Nelleman explores the motivations that have led Japanese multinational corporations to seek partnerships with international subsidiaries.

Advocating for the utility of technology innovation centers (TIC), Ebrahim Radhi demonstrates how these organizational frameworks can serve as hubs for the transmission of knowledge. In his chapter, Radhi describes a 2011 pilot project to hire a consulting company to research the potential for the domestic manufacture of products that had been previously been imported from abroad. Based on the successful outcome of this partnership, Radhi concludes that the establishment of locally based TICs would strengthen the manufacturing capabilities of local industry in Bahrain. Drawing on Qatar's successful adoption of a variety of TIC models, Radhi makes a persuasive case for how Bahrain can benefit from following a similar blueprint.

Gao Xudong's chapter analyzes the characteristics of Chinese oil and gas companies investing in the Gulf. Due to their relatively late entrance into the regional market, Chinese firms have had to bid for high risk projects. Gao argues that when these companies have successfully overseen the development of oil fields deemed too risky for other companies, this has helped them win additional contracts. Chinese companies have also demonstrated a track record for establishing close and mutually beneficial partnerships with host country firms. Finally, Chinese companies have demonstrated a capacity to undertake complex operations on their own, leading to greater efficiencies and cost benefits.

The final four chapters of the book are oriented toward examining the micro dynamics of communication and collaboration, or, in other words, the front lines of efforts to advance a knowledge-based economy in the Gulf. The first of these is an article by Sheikha Shamma bint Sultan bin Khalifa Al Nahyan and Laura L. Matherly that examines the factors that lead to effective knowledge transfer between expatriates and nationals. Based largely on 496 survey responses from professionals at public or semi public organizations, this is one of the few existing research studies to explore inter-professional dynamics at the ground level. Given the resources dedicated to nationalization campaigns across the Gulf, this is a particularly timely piece with important policy implications as it advances our understanding of the kinds of interactions that are likely to yield effective knowledge transfer.

Michael Corman and Jason Hickey tackle a similar topic, but use a different methodological strategy. Corman and Hickey's chapter adopts an ethnographic and phenomenological approach to investigating barriers to interprofessional collaboration in Qatar's healthcare system. Also, importantly, Corman and Hickey's paper is less focused on knowledge transmission between expats and nationals, and more focused on how professionals with vastly different professional socialization and work experiences navigate the challenges of working in a demanding and stressful profession. Based on interview data, Corman and Hickey document the communicative strategies that these professionals adopt in order to overcome the hierarchical barriers they encounter in the workplace; hierarchies that in some instances affect healthcare outcomes for patients.

Though Heba Nasser's paper is similarly focused on medical professionals, her emphasis is on the macro-level effects that healthcare worker migration has on sending countries such as Egypt, Lebanon, and Sudan and receiving countries in the GCC. Nasser's chapter provides a broad overview of regional migration flows, examining the push and pull forces that influence the decision to migration. Her chapter carefully investigates regional context for migration by looking at three national-level case studies: Egypt, Lebanon, and Sudan. Her paper concludes with a series of policy recommendations to promote sustainable medical professional migration between sending and receiving countries. These recommendations are designed to balance the continuing health care needs in the GCC, which she predicts will continue to increase, while not undermining health-care systems in sending countries such as Lebanon and Egypt.

Finally, Martin Lestra and Elyamine Settoul tackle a related subject in their chapter on the migration of French nationals of North African descent to the Gulf. Lestra and Settoul argue that for these primarily professional migrants, working in the Gulf serves two important objectives. First, as European nationals working in the Gulf, they are able to elude the discriminatory labor practices in France. In the GCC, they are able to acquire work experience that may eventually translate into more professional opportunities in France. Second, while working in the GCC, they are able to reestablish their cultural identities in a cosmopolitan environment that is culturally familiar and that also has a large North African population - an environment in which their identities as Muslim Arabs and French citizens are validated. In other words, what hindered their professional success in France, is now an asset. Within this context, Lestra and Settoul argue that these migrants possess a form of spatial capital, a combination of multicultural knowledge along with the international mobility granted through their status as French citizens, a resource that many have been unable to capitalize on until now. Relatedly, Lestra and Settoul show how GCC countries such as Qatar have actively recruited in areas of Paris with large North African diaspora populations that have a lengthy history of marginalization. In doing, they demonstrate that policies designed to promote a knowledge-generative economies are increasingly transnational in scope and not just domestically oriented.

Note

1 At its summit in Algiers in September 2016, OPEC decided to reduce oil production between 32.5 million barrels a day and 33 million barrels a day. Though this is a strategy to defend market price, the overall atmosphere is driven by profound mistrust among its members and thus favors the defense of market share.

References

Abi-Mershed, O. (ed) (2010) *Trajectories of education in the Arab world: legacies and challenges*, London; New York: Routledge.

Al-Atiqi IM, Al-Rashed AA & Ali FM (2009) "The contribution of private universities in higher education equity in Kuwait", *Toward an Arab higher education space: international challenges and societal responsibilities*, ed. B. Lamine, Beirut: United Nations Education, Scientific and Cultural Organization, 31 May-2 June: 111.

Al-Hamarneh, A. (2006) "German-Arab cooperation in education and science and the intercultural dialogue-an Arab perspective" in *Mediterranean Sea – gap or bridge? Perspectives on cooperation in education and scholarship between Germany and the Arab World*, ed. Herbert-Quandt-Stiftung, Bad Homburg: Societätsverlag, 118-122.

Al-Hatlani, I. (2016) 01/25-last update, *Gas Price Hike, Subsidy Cuts Shake Up Saudis* [Homepage of AlMonitor], [Online]. Available: http://tinyurl.com/ztfjnjc [2017, 01/10].

Al-Lamki, SA (2006) "The development of private higher education in the Sultanate of Oman: perception and analysis", *International Journal of Private Education*, vol. 1: 54-77.

Allen-Ebrahimian, B. (2016) 2 May-last update, *China Wants to Feed the World's 1.6 Billion Muslims* [Homepage of Foreign Policy], [Online]. Available: http://tinyurl.com/hfrtd5t [2017, 01/10].

Al-Mezaini, K. & Rickli, J. (eds) (2016) *The Gulf Small States: Foreign and Security Policies*, London: Routledge.

Altbach, PG & Salmi, J. (eds) (2011) *The road to academic excellence: The making of world-class research universities*, Washington, DC: The International Bank for Reconstruction and Development / The World Bank.

Andersen, LE & Jiang, Y. (2015) *Oil, Security, and Politics: Is China Challenging the US in the Persian Gulf?* Copenhagen: Danish Institute for International Studies.

Armitage, RL & Nye, JS, Jr (2007) *CSIS Commission on Smart Power: A smarter, more secure America*, Washington, DC: Center for International & Strategic Studies.

Beck, M. (2006) "Germany's contribution to building knowledge-based societies in the Arab World. How to match the aims of a civilian power with the goals of the Arab Human Development Reports" in *Mediterranean Sea – gap or bridge? Perspectives on cooperation in education and scholarship between Germany and the Arab World*, ed. Herbert-Quandt-Stiftung, Bad Homburg: Societätsverlag: 48-63.

Bertelsen, RG (2016) "The political context for transnational actor soft power: Classical American overseas missionary universities and the state" in: ed. N. Chitty, Beijing: Communication University of China.

Bertelsen, RG (2014a) "American missionary universities in China and the Middle East and American philanthropy: Interacting soft power of transnational actors", *Global Society,* vol. 28, no. 1 (Special Issue: American Philanthropy and the Hard, Smart and Soft Power of the United States): 113-127.

Bertelsen RG (2014b) "The university as a transnational actor with transnational power: American missionary universities in the Middle East and China", *PS: Political Science & Politics,* no. July: 624-627.

Bertelsen, RG (2012a) "The Effect of Public and Private Decisions on University Governance on the Transnational Relations of American-Associated Universities in the Middle East", *Revue des mondes musulmans et de la Méditerannée,* vol. 131, no. 1: 45-62.

Bertelsen, RG (2012b) "Private foreign-affiliated universities, the state and soft power: the American University of Beirut and the American University in Cairo", *Foreign Policy Analysis,* vol. 8, no. 3: 293-311.

Bertelsen RG (2009a) *American- and French-affiliated universities in the Middle East as 'Information and Resource Bridges' to the West,* The Dubai Initiative, Belfer Center for Science and International Affairs, Cambridge, MA: Harvard Kennedy School of Government.

Bertelsen RG (2009b) *Private higher education in the GCC: Best practices in governance, quality assurance and funding*, The Dubai Initiative, Belfer Center for Science and International Affairs, Cambridge, MA: Harvard Kennedy School of Government.

Bertelsen, RG *Public Policy, University Governance and Transnational Linkages: American-Associated Universities in the Middle East,* Unpublished paper edn.

Bertelsen, RG, Dai, Y. and Solinap, G (2013) ""Learning, when you are not learning": artistic, scientific, professional and political culture at leading universities in Britain, France, USA and China" in *Teaching and learning culture, negotiating context,* eds. X. Du, A.A. Jensen & M.J. Kirkebæk, Rotterdam: Sense Publisher: 145-162.

Bertelsen, RG, Du, X. and Søndergaard, MK (2014) "Sino-Danish brain circulation: scholarship, capacity and policy", *International Journal of Business and Globalisation,* vol. 12, no. 2: 142-160.

Bertelsen, RG, Du, X. and Søndergaard, MK (2013) "Strategic knowledge collaboration between Danish business and Chinese academia", *Journal of China and International Relations,* vol. 1, no. 1: 66-87.

Bertelsen, RG and Noori, N *The new medievalism and transnational governance of higher education institutions: A comparison of American and European accreditation regimes in the Middle East,* Unpublished paper edn.

Carter, J. (1980) 23 January 1983-last update, *The State of the Union Address Delivered Before a Joint Session of the Congress, 23 January 1983.* Available: http://www.presidency.ucsb.edu/ws/?pid=33079 [2017, 01/10].

Chandrasekaran, R. (2014) 9 November-last update, *In the UAE, the United States has a Quiet, Potent Ally Nicknamed "Little Sparta"* [Homepage of Washington Post], [Online]. Available: http://tinyurl.com/zw5zpbq [2017, 01/10].

Council on Foreign Relations (2007) *American Universities in the Middle East: Agents of Change in the Arab World,* Council on Foreign Relations, www.cfr.org.

Daojiong Z. and Meidan M. (2015) *China and the Middle East in the New Energy Landscape*, London: Chatham House.

Davidson, C. and Smith, PM (eds) (2008) *Higher education in the Gulf states: shaping economies, politics and cultures,* London: Al-Saqi Books in association with The London Middle East Institute at SOAS.

Diaa, S. (2016) 02/22-last update, *GCC states in agreement on 5% VAT* [Homepage of Gulf News], [Online]. Available: http://tinyurl.com/jh4sal4 [2017, 01/10].

Djerejian, EP (2003) *Changing minds winning peace: a new strategic direction for U.S. public diplomacy in the Arab & Muslim world. Report of the Advisory Group on Public Diplomacy for the Arab and Muslim World,* Washington, DC: The Advisory Group on Public Diplomacy for the Arab and Muslim World.

Haddad-Fonda, K. (2016) 11 May-last update, *China's Massive, Garish Theme Park for the Muslim World* [Homepage of Foreign Policy], [Online]. Available: http://tinyurl.com/h2ancx7 [2017, 01/10].

Harris, M. (2015) 14 August-last update, *Documents confirm Apple is building self-driving car* [Homepage of The Guardian], [Online]. Available: http://tinyurl.com/j6dfbdq [2017, 01/10].

Herbert Quandt Stiftung (ed) (2006) *Mediterranean Sea – gap or bridge? Perspectives on cooperation in education and scholarship between Germany and the Arab World*, Bad Homburg v. d. Höhe: Herbert Quandt Stiftung.

Hvidt, M. (2015) "The State and the Knowledge Economy in the Gulf: Structural and Motivational Challenges", *The Muslim World*, vol. 105, no. 1: 24-45.

Kamola I. and Noori N. (2014) "Higher education and world politics", *PS: Political Science & Politics*, no. July: 599-603.

Kamrava, M. (2015) *Qatar: small state, big politics*, Ithaca, NY: Cornell University Press.

Kuznetsov, Y. (ed) (2006) *Diaspora Networks and the International Migration of Skills: How Countries Can Draw on Their Talent Abroad*, Washington, DC: The World Bank.

Lane, JE (2010) *International branch campuses, free zones, and quality assurance: policy issues for Dubai and the UAE*, Dubai: Dubai School of Government.

Lewin, T. (2009) *George Mason University, among first with an Emirates branch, is pulling out.*

Lord, C. (2006) *Losing hearts and minds? public diplomacy and strategic influence in the age of terror*, Westport, CT: Praeger Security International.

Mazawi, AE and Sultana, RG (eds) (2010) *Education and the Arab 'world': political projects, struggles, and geometries of power*, New York; London: Routledge.

Ministry of Higher Education, Sultanate of Oman (nd) *Private universities and college*s [Homepage of Ministry of Higher Education. Sultanate of Oman], [Online]. Available: http://tinyurl.com/zp5s2ur [2017, 10 January].

Noori, N. (2014) "Does academic freedom globalize? The diffusion of the American model of education to the Middle East and academic freedom", *PS: Political Science & Politics*, no. July: 608-611.

Noori, N. and Anderson, P. (2013) "Globalization, governance, and the diffusion of the American model of education: Accreditation agencies and American-style universities in the Middle East", *International Journal of Culture, Politics, and Society*, vol. 26, no. 2: 159-172.

OECD (2008) *The global competion for talent: Mobility of the highly skilled*, Paris: Organisation for Economic Cooperation and Development.

OECD (2002) *Policy Brief: International mobility of the highly skille*d, Paris: Organisation for Economic Development and Cooperation.

Office Méditerranéen de la Jeunesse (nd) *Catalogue de formation de niveau Master 2 et Doctorat labellisées OMJ* [Homepage of Office Méditerranéen de la Jeunesse], [Online]. Available: http://catalogue.mediterraneanofficeforyouth.com [2017, 01/10].

Olson, MJ (2009) "Liberal arts education as a commodity in GCC countries and e-learning's implications", *3rd Liberal Arts Conference proceedings*, Salmiya, Kuwait, May.

Olson, MJ (2008a) "Resolving Liberal Arts' Needs for a Muslim Majority Setting: Some Organizational Issues for Research", *The Global Studies Journal*, vol. 1, no. 2: 55-62.

Olson, MJ (2008b) "When wasta and liberal arts conflict: a case study of a private university", *AUK occasional papers: liberal arts & business series*, vol. 2: 24-32.

Peled, AR (2012) *Adaptation, ideals, vision and deals: the rise of American higher education in the Arab world*, Cambridge, MA: Harvard University Press.

Powell, WW and Snellman, K. (2004) "The knowledge economy", *Annual Review of Sociology*, vol. 30: 199-220.

Rice, C. (2006) 5 January 2006-last updat*e*, *Remarks at the U.S. university presidents summit on international education dinner.* Washington DC. Available: http://www.state.gov/secretary/rm/2006/58750.htm [2017, 01/10].

Rickli, J. (2016a) "New Alliances Dynamics and their Impact on Small GCC States", *Third World Thematic*, vol. 1, no. 1: 1-19.

Rickli, J. (2016b) "The Political Rationale and Implications of the United Arab Emirates' Military Involvement in Libya" in *Political Rationale and International Consequences of the War in Libya*, eds. D. Henriksen & A.K. Larssen, Oxford: Oxford University Press: 134-154.

Rickli, J. (2014) "Clean Energy as a Niche Strategy for Small States to Guarantee Energy Security. The Example of the Gulf Countries" in *Political Economy of Energy Reform: The Clean Energy-Fossil Fuel Balance in the Gulf*, eds. G. Luciani & R. Ferroukhi, Berlin: Gerlach Press: 265-288.

Rickli, J. and Charles, L. (2013a) "The UAE's Energy Dilemma", *Capital Business*, vol. 8, no. 4: 32-35.

Rickli, J. and Charles, L. (2013b) "The UAE's Sovereign Wealth Funds- Securing the future?", *Capital Business*, vol. 8, no. November: 32-35.

Rugh, WA (2006) *American encounters with Arabs: the "soft power" of U.S. public diplomacy in the Middle East*, Westport, CT: Praeger International Security.

Saxenian, A. (2006) *The new argonauts: regional advantage in a global economy*, Cambridge, MA: Harvard University Press.

Saxenian, A. (2005) "From brain drain to brain circulation: transnational communities and regional upgrading in India and China", *Studies in Comparative International Development*, vol. 40, no. 2: 35-61.

Saxenian, A. and Sabel, C. (2008) "Roepke Lecture in Economic Geography Venture Capital in the 'Periphery': The New Argonauts, Global Search, and Local Institution Building", *Economic Geography*, vol. 84, no. 4: 379-394.

Seguin, B., Singer, PA and Daar, AS (2006) "Scientific diasporas: scientists from the developing world who work in more developed countries are often underutilized resources and should be cultivated for the benefit of their countries of origin", *Science*, vol. 312, no. 5780: 1602.

Stark, D. (2011) *The sense of dissonance: Accounts of worth in economic life*, Princeton, NJ: Princeton University Press.

The Middle East Institute 2010, *Higher education and the Middle East: serving the knowledge-based economy*, Washington, DC: The Middle East Institute.

The World Bank Group (2009) 2009-03-01-last update, *Education for the Knowledge Economy* [Homepage of The World Bank Group], [Online]. Available: http://go.worldbank.org/I8T7C0VPV0 [2017, 01/10].

Thomson, SC (2004) *Teachers give bad grades to U. of Missouri-linked school in Kuwait*.

United Nations Development Programme (2003) *Arab human development report 2003: building a knowledge society*, Amman.

World Bank (2008) *The road not travelled: education reform in the Middle East*, Washington, DC.

Young, KE (2016) 04/06-last update, *Drop in the Bucket: Reduced Fuel Subsidies Offer Little Deficit Relief* [Homepage of The Arab Gulf States Institute in Washington], [Online]. Available: http://tinyurl.com/hrtgf4r [2017, 01/10].

Zweig, D., Changgui, C. and Rosen, S. (2004) "Globalization and Transnational Human Capital: Overseas and Returnee Scholars to China", *The China Quarterly*, vol. 179, no. September: 735-757.

2

The Role of Technology Entrepreneurship in Promoting Sustainable Development in the GCC

Daniel Goldman

Introduction

Traditional paradigms in development have viewed advanced technology as the culminating phase of industrialization, modernization, and economic diversification, rather than as a means to increase long-term growth and competitiveness in countries confronting significant economic challenges. The reliance of the GCC states on petroleum exports has left national fiscal and monetary policies highly vulnerable to price shocks, which commodity stabilization funds and sovereign wealth funds have ameliorated only in part, and inhibited the development of a robust private sector.

There are very few historical instances of deliberate national transitions to advanced technology across multiple sectors prior to the advent of widespread private sector development. In each case, it was a strategy of last resort pursued by countries, such as post-war South Korea, post-independence Singapore, and, more recently, post-conflict Rwanda, lacking natural resources, friendly neighbors, political stability, or an existing industrial base. In each such case at the urban level, such as post-Cultural Revolution Shenzhen and 1990's Hyderabad, they were experimental strategies selected by cities lacking access to natural resources and stagnating economically. Despite these extraordinary successes in long-term economic development, they are frequently dismissed as anomalies, and their early and rapid transition to advanced technology is largely ignored, rather than proffered as a model for regions whose natural resources, political stability, and trade relationships enable them to stave off short and medium-term social unrest, environmental degradation, and economic decline.

The initiatives to transition to post-carbon economies in the GCC can be realized most rapidly, effectively, and securely through the promotion of high-technology industries across sectors; these include bioinformatics, nanotechnology, genomics, and solar and geothermal energy. The widespread funding of academic and private research laboratories, strengthening of intellectual property rights, extension of critical regulatory, financial, and tax incentives to technology entrepreneurs, and international recruitment of top-flight scientists and technologists will foster increased international investment in the region, accelerate the development of a high-skilled labor force, increase knowledge transfer and collaboration between GCC universities and their international counterparts, advance efforts towards environmental conservation, and promote national economic self-sufficiency and sustainability.

Literature Review

In recent years, academics, policy makers, and multilateral institutions have increasingly stressed the need to emphasize sustainability in promoting development in the GCC. They have acknowledged the negative environmental effects of reliance on extractive industries (Sale and Feary, 2011), as well as the need for increased diversification in exports to increase economic stability and promote job creation. They have also promoted transitioning to knowledge-based economies as a means to reduce the reliance on extractive industries and increase human capital in the region (Callen et al., 2014; Brinkley et al., 2012; Economic Vision 2030 Bahrain; UAE Vision 2021).

Numerous proposals advocate a combination of industrialization (Mueller, 2012) and the replication of strategies pursued by countries that successfully diversified their economies away from a focus on oil as a means to achieve economic diversification. In a recent working paper, Fuad Hasanov and Reda Cherif (2014) of the IMF contend that the Gulf countries should emulate the diversification strategies of Indonesia, Malaysia, and Mexico, and cite the relative and absolute decline in export sophistication in the GCC as a cause for slowed economic growth. They note that technology parks and industrial clusters have, thus far, yielded few results in diversification and technology development. Allam Ahmed and Ibrahim Alfaki (2013) argue for the importance of science, technology and innovation in promoting growth in the UAE, and note the developmental arc of Singapore and South Korea as relevant reference points for the Emirates.

Other recent proposals emphasize targeted policy interventions to promote economic growth and financial stability. A recent report from Natalija Gelvanovska, Michel Rogy, and Carlo Maria Rossotto (2012) of the World Bank emphasize the need to improve broadband connectivity to enhance integration into the global economy and increase economic diversification. Martin Hvidt (2011, 2015) notes that the oil-driven regional economic model throughout the region and the real estate and bank-driven economic model of Dubai have dampened societal incentives to emphasize education and innovation, and argues for the reform of incentive structures to promote a transition to a regional knowledge economy. A recent paper from Zsofia Arvai, Ananthakrishnan Prasad and Kentaro Katayama (2014) of the IMF focuses on the use of macroprudential tools to promote national and regional financial and economic stability in the face of oil price volatility.

The literature on sustainable development is extensive, with several contributions in recent years regarding regional challenges (Sillitoe, 2014: Cevik, 2014). Amer Al-Roubaie and Jamal Al-Zayer (2006) emphasize the role of technology transfer in promoting regional sustainability. Jeremy Hall, Gregory Daneke, and Michael Lenox (2010) advance a skeptical perspective on sustainable development and entrepreneurship, contending that economic growth and sustainability are often incompatible and asserting that the mechanisms by which entrepreneurship could advance sustainable development are inadequately understood. Dwaa Osman (2015) contends that rentier state dynamics and the lack of a systemic approach to promoting innovation have inhibited regional growth. Crystal Ennis (2015) argues that attempts to promote entrepreneurship in Qatar and Oman have been excessively top-down and failed to address underlying structural issues related to governmental patronage and labor market inefficiencies.

Potential Models

In 1961, South Korea was one of the poorest countries in the world, with an annual per capita GDP of $82, 13% unemployment, and rampant inflation (Harvie and Lee, 2003: 258). The population and its national infrastructure had been devastated during the Korean War; mass migration from the North had created overcrowding, while the country lacked a significant industrial base or natural resource endowments (Harvie and Lee, 2003: 267-68). The government's initial emphasis on export-led development and the promotion of chemical and heavy industries generated significant growth, but ultimately led to stagnation and a reversion to high inflation by the end of the 1970's, while artificial strengthening of the currency and underinvestment in technology and human capital led to high trade imbalances (Harvie and Lee, 2003: 272).

During the next phase of development, although the country was still relatively poor, the government emphasized the promotion of small and medium enterprises and technology, which including emphasizing information and communication technology strongly throughout the educational system (Sanchez et al., 2011), and shifted towards a focus on bottom-up private sector-led development (Harvie and Lee, 2002: 273). One mechanism used to promote technology entrepreneurship was through innovative financing mechanisms, including the Technology Development Assistance Funds, to promote R&D and intellectual property protection; by 2010, the South Korean government spent near $1 billion annually on technology R&D for SMEs, while the total national expenditure on R&D for SMEs alone in 2010, $6.7 billion, exceeded all R&D spending for all the GCC countries combined.

South Korea has intensified its commitment to advanced technology in recent years; it currently ranks first in the world on Bloomberg's Global Innovation Index, with important current work in fields such as nanotechnology, space science, and stem cell research, and leads the world in R&D spending as a percentage of GDP (*Nature*, 2015). While it still remains well behind many GCC countries on aggregate measures of wealth, it has caught up to high-income countries more rapidly than anyone had anticipated. The country has sustained extraordinary economic growth under this program, as well as a dramatic rise in social indicators related to standard of living. It currently has an unemployment rate of 2.7%, among the lowest globally, and ranks fifteenth in the world on the UN's Human Development Index, ahead of Japan and France. The South Korean government's emphasis on the promotion of sustainable economic development through technology R&D and the advancement of the private sector, with a significant role for entrepreneurs and small businesses, played a significant role in contributing to national growth.

In more recent years, Rwanda successfully generated technology-led economic development after one of the most horrific and tragic conflicts in modern times. In the aftermath of the 1994 genocide, in which one sixth of the population was killed, the country faced extremely high unemployment, abject poverty, rampant inflation, and hostile relations with its neighbors. It lacked an existing industrial or manufacturing base and was largely bereft of natural resources (Gatete, 2015). There was a widespread reluctance from multilateral and bilateral donors to provide the country foreign aid (Kagame, 2010). Romain Murenzi and Michael Hughes (2006), who have worked for the Rwandan government on national science and technology policy, have framed the strategy in confronting these challenges as follows:

> Competitive advantage is partly based on an uneven endowment of resources around the globe, and partly on specialization, skilled labor, and the accumulated capital equipment that can be used to add value to natural resources. In the case of small African economies such as Rwanda, there is a shortage of natural mineral resources as well as land. Such economies have little alternative but to commit themselves to building their science and technology infrastructure and knowledge base as a foundation for their competitive advantage.

Recognizing that it had few other available options, the government emphasized technology research and development as a means to promote economic growth. The government understood that universities would need to play a critical role in this development, and grew the national university population from one in 1994 to fourteen in 2006, with many of the new institutions, such as the Kigali Institute of Science and Technology and the Kigali Health Institute, emphasizing basic research and technology development. At the same time, it integrated scientific and mathematical education more strongly into the curriculum at primary and secondary educational institutions (Murenzi and Hughes, 2006: 257-62).

While technology transfer constituted a key component of Rwanda's strategy, the government sought to emphasize the development of technologies that would address national and regional challenges. These included materials for sustainable construction in the real estate sector, telemedicine and diagnostics for malaria and HIV in the health sector, and novel rainwater harvesting and irrigation techniques in the agricultural sector (Murenzi and Hughes, 2006).

The Rwandan government emphasized the promotion of entrepreneurship, private sector-led development, and basic research as key components of its long-term strategy. President Paul Kagame noted in 2010:

> In the old Rwanda, everyone looked for a job in government because of the benefits and security. But nowadays they are thinking that the private sector holds the promise of a better life for their families and themselves.

The country's economic programs have been remarkably successful; it has achieved 8% annualized GDP growth, one of the highest rates in the world, and was ranked second in Africa in the 2014 World Bank Doing Business Report, which also cited it as the most improved economy globally since 2005. Over 5,000 km of optic fiber has been installed, giving Rwanda the fastest broadband access on the continent, and over 65% of the population has mobile phone access, up from a negligible amount ten years prior (Rwandan Economic Development Board, 2014). While Rwanda still faces widespread poverty and confronts significant challenges, its progress based on heavy investment in science and technology infrastructure, education, and industry and the promotion of entrepreneurship may be relevant for understanding the prospective development of countries that have many advantages that Rwanda lacks.

There are also several examples of successful transitions to high technology prior to industrialization at the regional or urban level, including Hyderabad, India in the 1990's. While the city was widely viewed as economically uncompetitive at the beginning of the period, government policies had a substantial impact on its rapid growth; by the end of the decade, government officials from Mumbai sought to emulate its success. The number of software firms in the city grew from a negligible amount in 1992 to over one thousand in 2000, while export revenues increased from less than $1 million annually to near $125 million annually over the same period (Biswas, 2004).

The government stimulated the growth of the software sector by creating hardware parks alongside software parks and developing other IT-intensive sectors such as biotechnology, out of recognition that each technological subspecialty likely would be beneficial to the development of the other. Despite the poverty of the region, the state government created strong international satellite links and installed an advanced fiber optic network that it encouraged the private sector to expand. In addition, the government created a new college, the Indian Institute of Information Technology, as well as creating distance learning initiatives focused on sector-related education. It also transitioned its own platforms online through an e-governance program, and offered incentives for infrastructure, real estate, and power specific to the technology sector (Biswas, 2004).

The development of Shenzhen offers another relevant example; since its designation as a Special Economic Zone in 1980, it has evolved from a small fishing village to a city of over 10 million people that has attracted over $30 billion in foreign direct investment and annually exports billions of dollars of goods and services internationally. While numerous attempts to replicate these zones in other environments have met with mixed success (Norman, 2014), they often have coupled measures such as free trade and undifferentiated tax incentives with top-down national or provincial control over the zones, rather than emphasizing accelerated technology development, as Shenzhen did from the outset.

The local government has prioritized advancing its economy as well as national technological capacity through the promotion of research and development in information technology, the biological sciences, and new energy, and has allocated $1.69 billion for investment in these core areas since 2009 (Sharif and Tang, 2014). The city received 62,000 invention patents in 2013, second in China and an order of magnitude more than the GCC countries combined, and spent 4% of its GDP on R&D that year, highest in China, an order of magnitude higher than any GCC country, and more on a percentage basis than any country other than South Korea (WIPO, 2015; Shenzhen Government Online, 2014; also see Chen & Kenney 2007 for a more skeptical view). It is also emphasizing scientific, technological, and economic integration and cooperation with Hong Kong and Macau, establishing local graduate universities and research emphasizing technological education, including the National Supercomputing Center, the Kuang-Chi Institute of Advanced Technology, and the Shenzhen Institute of Advanced Technology of CAS (Shenzhen Government Online, 2014), and fostering collaboration among universities, research institutes, and the private sector. Like Hyderabad, the city has focused on developing competitive high-tech educational institutions, corporations, and professionals locally and training local rather than importing Western workers and brands. In addition, it is focusing on incubating small and medium enterprises likely to promote technological innovation (Sharif and Tang, 2014).

Each of the cases addressed above – South Korea, Rwanda, Hyderabad, and Shenzhen – reflects a strategy of technology development by necessity, given many of the following factors: widespread poverty, lack of adequate land, paucity of natural resources, post-conflict or post-colonial environment, lack of regional trading partners, and low level of industrialization. These countries and cities pursued advanced technology as a last resort, as traditional thinking dictated that advanced technology should be the culminating phase of economic development.

GCC Developmental Challenges

The GCC countries, in contrast, have significant wealth, substantial natural resource endowments, relatively stable governments with little recent history of conflict, and numerous potential trading partners, although they have low levels of industrialization and varying levels of land availability. While other forward-looking accounts of GCC growth have focused principally on following the developmental path of other countries with substantial natural resource endowments, and found few normative models, the GCC countries might instead capitalize most effectively on their advantages through resource conversation and the promotion of a robust private sector focused on technology entrepreneurship.

While commonly used macroeconomic indicators, such as GDP and GNI per capita, suggest that the GCC countries are performing well, an examination of the underlying data reflects that these countries confront significant economic challenges. The vast majority of regional exports, as well as a significant proportion of regional GDP, result from the liquidation of assets – or resource extraction – rather than economic value creation. Orthodox schools of economics recognize value only when a transaction occurs, and thus treat existing stores of value that have not been traded as worthless. As a result, while a corporate executive who repeatedly liquidated billions of dollars worth of a commodity for a price substantially below fair market value likely would be fired, a national government official who did so probably would be praised for raising the country's GDP. The economist Dieter Helm (2014) has framed the challenge as follows:

> [N]atural capital is typically treated as if it has zero value. It is typically left out of our accounts – national, corporate, and for public bodies. There is little recognition of critical thresholds, few metrics to evaluate what is damaged, and an absence of a register of assets at risk... The accounting problem is worse than simply the exclusion of natural assets. There are no comprehensive national balance sheets at all. National income, expenditure, and production are all flows, not stocks... GDP is a flow concept. It takes no account of any asset depreciation, national or man-made, because there are no assets in its calculation.

This is a particularly acute regional challenge; according to the most recent statistics, Oman, Saudi Arabia, Qatar, and Kuwait had among the highest global percentages of income derived from natural resource depletion (World Bank, 2011). The adoption of regional natural capital accounts would lead to a broader recognition that the extent of current resource depletion is not only economically unsustainable, but inefficient.

To mitigate short-term effects on competitiveness, multilateral bodies should accelerate the international adoption of these accounts. While the UN and World Bank have initiated efforts to coordinate the standardization of natural capital records and assist countries with their voluntary collection (World Bank, 2015), these institutions should transition to requiring their adoption at the national level by 2025 and incorporating them into macroeconomic indicators, while providing the necessary technical assistance and capacity building to facilitate an effective international process.

The reliance on fossil fuels to generate economic growth also contributes to overconcentration in sources of government revenue. While many other high-income governments rely on a variety of sources of revenue, and depend on revenues from a robust, highly diversified private sector to insulate themselves from sector-specific shocks, each GCC countries depended on petroleum for over three fourths of its revenue, as of 2011 (Hvidt, 2013); in Saudi Arabia, petroleum accounts for over 90% of government revenues (Ben-Rubeian, 2015).

The regional dependence on petroleum as a source of government revenue, coupled with high and unpredictable volatility in oil prices, creates numerous challenges for fiscal policy, which the establishment of sovereign wealth funds and stabilization funds have ameliorated only in part. The capacity to budget, plan, forecast, and allocate capital are inhibited substantially by the high volatility in government revenues. While others have argued in favor of regional diversification towards sectors related to oil, this strategy would still leave GCC countries quite vulnerable to underlying commodity price fluctuations. An approach to diversification centered around assets largely uncorrelated with oil prices would provide a mechanism to smooth out government spending and facilitate the establishment of more effective and long-term fiscal policy.

The regional reliance on fossil fuels has also contributed to market distortions, inefficiencies, and environmental damage (Beutel, 2012). The IEA and IMF have estimated that the GCC countries provided over $100 billion annually in fossil fuel subsidies in 2011 (Charles et al., 2014), while a more recent paper that accounts more completely for negative externalities suggests that the figure could be substantially higher (Coady et al., 2015).

These subsidies have numerous negative effects, including increased fiscal burdens, reduced incentives for renewable energy development and usage, higher carbon emissions, increased transportation congestion, and excessive levels of pollution (Coady et al., 2015; Charles et al., 2014). As the subsidies primarily accrue to wealthy households, they are an ineffective mechanism for poverty reduction (Coady, 2015: 5). The recent sharp drop in oil prices creates a significant opportunity for governments, including those in the GCC, substantially to reduce fossil fuel subsidies without political repercussions (Lagarde, 2015); although the countries have recently done so to an extent (c.f. e.g. Kerr and Clark, 2015), each GCC country likely would benefit from further reductions. The subsidies could be reallocated to stimulate the growth of sectors that would provide economic diversification and use advanced technologies to address regional economic, social, and environmental challenges.

The regional reliance on fossil fuels has inhibited the development of a robust private sector in the region, while rentier state dynamics have reduced the incentives for national governments to promote advanced education, technology development, and regulatory reforms. While regional countries rank quite highly on World Economic Forum basic measures of economic competitiveness, all except the UAE place well below their income ranks and high-income country averages on higher education and training; on innovation, Bahrain and Oman rank below Azerbaijan and Guyana, while Kuwait ranks behind Burkina Faso and Mongolia (Schwab and Sala-I-Martin, 2015). On Ricardo Hausmann and Cesar Hidalgo's economic complexity rankings, which measure the sophistication of national outputs, the UAE and Saudi Arabia rank 57th and 58th respectively, below El Salvador; Kuwait and Oman rank 75th and 76th respectively, below Cuba and Vietnam; and Qatar 81st, below Namibia (Hausmann and Hidalgo, 2011).

Regional countries also fall well below their high-income global counterparts on the World Bank's Doing Business Indicators, particularly for those measures most relevant to entrepreneurship. All regional countries but the UAE rank below 100th on ease of starting a business, while all but Bahrain rank in the bottom half globally on resolving insolvency, and all but Saudi Arabia rank in the bottom half globally on getting credit.

On protection of minority investors, particularly critical to venture capitalists, angel investors, and incubators, Bahrain ranks 104th and Oman and Qatar are tied for 122nd, while no Gulf country ranks higher than 108th globally on enforcement of contracts. On overall rankings, Rwanda, which is an order of magnitude poorer than any regional country, ranks higher than five of the six Gulf countries. Regulatory reforms to comport with international best practices likely would prove materially helpful to catalyzing regional entrepreneurship.

Regional financial markets also remain underdeveloped relative to the size of the economies. As of 2012, regional equity markets and overall financial markets each comprised less than 1% of global equity and financial markets, respectively (Kern, 2012; World Bank, 2015), and reflected less than half the region's contribution to global output. The strengthening of regional financial markets would have a catalytic effect on the entrepreneurial ecosystem by providing enhanced exit options for early-stage investors.

The countries also rank poorly on numerous environmental indicators. Every country in the region ranks among the eleven highest per capita global emitters of carbon dioxide and the thirteen highest per capita users of energy, although none are in the top seventy-five of GDP per unit of energy usage, reflecting high levels of inefficiency and waste in the energy system (World Bank, 2015). Each country in the region produces less than 0.5% of its energy and electricity from renewable sources (United Nations, 2015). The UAE, Saudi Arabia, Oman, and Kuwait rank among the fifteen countries in the world with the highest concentrations of harmful particulate matter, and are the only high-income countries on that list (UNEP, 2015).

Despite the negative long-term economic and environmental effects of a reliance on fossil fuel extraction, and a stated vision by each country to promote a transition to a knowledge economy, each of the GCC states remains heavily dependent on petroleum for economic growth and revenue generation. The strength of local currencies, high regional costs of living, and high level of competition globally due to widespread industrialization in low-income countries make a transition to manufacturing relatively unappealing economically –

particularly for low and medium-value add goods – while the negative environmental effects of manufacturing make it unattractive from an ecological perspective. A shift to focusing on research and development-intensive, high-technology, high-value add goods and services, including those that address national social and environmental challenges, represents the region's most compelling strategy for promoting long-term sustainability.

Knowledge Transfer and Researcher/Professional Mobility

While several Gulf governments recently have announced significant initiatives towards technology development, including the UAE's plan to launch a Mars mission by 2021 (Guardian, 2015) and its development of the carbon-neutral Masdar City, these efforts have, thus far, proven inadequate. As measured by key indices of innovation, including technology startups founded, sold, and floated in IPOs; patents granted; scientific and technical publications; scientific awards and prizes; economic contribution of technology to GDP; university and research laboratory technology transfer; and R&D spending as a proportion of GDP, the Gulf states still lag well behind their high-income-country global counterparts, and have failed to improve significantly compared to their peers since national plans were launched promoting technology development. To the extent economic diversification away from fossil-fuel related industries has occurred, it has been into sectors such as real estate, banking, insurance, and others based primarily on capital accumulation and financial engineering rather than knowledge accumulation and transfer.

The Gulf states can harness their core strengths effectively to promote technology entrepreneurship and development through bottom-up knowledge transfer, as well as through the promotion of researcher and professional mobility in high-technology sectors. Initial efforts to attract educational institutions and technology corporations have focused on importing Western brand-name institutions, often those whose innovation is a generation behind, rather than, for instance, Taiwanese or Indian institutions that have flourished with fewer resources.

We have discussed above the challenges posed by the reliance of the GCC states on the exportation of fossil fuels; their emphasis on the importation of Western technology is likewise problematic. Solutions optimized for a particular environment are often poorly suited for others, while challenges that occasion the development of technology frequently diverge sharply across regions. Omani efforts to promote solar energy, as one diplomat has noted, relied on solar panels designed and manufactured in Germany; when placed in the desert, they accumulated sand covers and failed to capture significant heat (Al-Shuaibi, 2015). Membranes used for energy-efficient reverse osmosis technology have not yet been adequately customized for high-salinity Gulf seawater; similarly, operating plants developed in the West for the technology function inefficiently in the heat, leaving GCC states to rely extensively on energy-intensive thermal desalination (Wali, 2014).

Rather than focusing on importing Western institutions and inventions, the Gulf states could more effectively promote technological innovation by promoting high-tech labor mobility and facilitating the free transnational flow of ideas, knowledge, and know-how among scientists, researchers, and technologists as a means to building their own indigenous world-class public and private-sector technology institutions. The promotion of conservation and a thoroughgoing program of economic diversification in the Gulf will require professionals with advanced knowledge in economics and accounting, while the promotion of technology entrepreneurship will benefit materially from professionals with expertise in cutting-edge fields such as artificial intelligence, renewable energy, genomics, and bioinformatics, in addition to venture capitalists, angel investors, and lawyers with expertise in technology startups.

Instead of emphasizing technology talent from the West, GCC countries ought to consider scientists and engineers from their own broader region. While the tragic recent forced migration in the broader region has occasioned much public debate and hand-wringing regarding responsibility for the care of the millions of refugees, the issue has been framed throughout as a humanitarian one, by the left, and a security one, by the right, in which the migrants are beneficiaries rather than agents in their own destiny. The United States benefitted materially from its welcoming policy towards refugee scientists from each side of World War II, who played instrumental roles in extraordinary technological ventures such as the Manhattan Project and the Apollo Program. The GCC could stand to gain significantly from adopting similar policies towards scientists, engineers, and researchers from war-torn countries in the broader region, to say nothing of the humanitarian benefits.

Saxenian (2006) and Zweig (1997) have chronicled how top Asian students, scientists, and researchers developed technological expertise from studying and working in hubs such as Silicon Valley and returned home to develop world-class native institutions. These researchers have not stressed the extent to which a deep knowledge of local, regional, and national context enabled these entrepreneurs, scientists, and technology executives to tailor solutions to home markets. The GCC government can promote knowledge transfer and mobility by redirecting fossil fuel subsidies and high public sector wages to funding for scholarships and fellowships for its own scientists and researchers abroad, as well as for international scientists to move to the region; providing further funding to technology transfer offices, incubators, accelerators, and venture capital firms, to accelerate innovation; harmonizing its intellectual property laws with best-in-class international standards; and developing other policy measures likely to appeal to R&D intensive startups, including the provision of net operating loss carryovers, favorable tax treatment for startup investors, and immigration and naturalization policies, similar to those developed in Singapore (Pereira, 2008), favoring high-quality scientists, researchers, and other high-tech professionals in particular areas of need.

In relying principally on top-down capital transfer rather than bottom-up knowledge transfer to promote technology development, the GCC governments have sought to replicate their successes in traditional industries such as oil, gas, and financial services, rather than recognizing the difficulties in precisely engineering, budgeting for, and declaiming technological innovation in advance. By promoting knowledge and technology transfer through policy measures promoting transnational researcher and professional mobility as well as bottom-up collaboration, the governments can transition effectively to an economic model focused on the conservation of natural resources and the commercialization and exportation of complex, high value-add, information-intensive services and goods, fostering a regional hub for high-technology products tailored to the particular energy, water, communication, healthcare, security, and telecommunications needs of their people as well as those of the broader Middle East region.

Prospective Technologies

> Tracing the history of development in the modern age reveals that obtaining scientific and technological knowledge coupled with the ability to use it have become among the critical strategic factors in the economic performance of various countries, particularly with the increase of globalisation and economic competition. (UNDP, 2014)

We now turn to specific technological areas that would benefit most from support as a means to promote sustainable development. A first is renewable energy; given the sharply increasing levels of regional energy demand, the current heavy dependence on fossil fuels, the suitability of the climate, and the match between seasonal electricity demand and solar availability (El-Katiri and Husain, 2014), it is a particularly compelling sector for increased investment. Although regional governments have set forward-looking targets for renewable energy usage and begun to ramp up R&D in this sector, the targets are non-binding, usage and production remain negligible, and high subsidies to traditional energy sources render the prospect of competition challenging (Haug, 2014).

Kuwait, through its Institute for Scientific Research, launched studies on solar energy as early as 1967, although it concluded them in 1986 with a determination that it was not cost-effective (Haug, 2014). This determination, however, relied on less advanced and far more expensive renewables technologies, coupled with far lower oil prices. More recent studies by Bloomberg New Energy Finance and the Emirates Solar Industry Association suggest that in certain countries in the region, these technologies are now nearly price-competitive with oil and natural gas, even before the effects of fossil fuel subsidies are taken into account (Haug, 2014).

Geothermal energy is another promising area of R&D, as it emits 1% of the carbon dioxide of oil when used and requires a tenth as much land per unit of energy as solar panels. This is particularly the case in Saudi Arabia, where geological formations enable the efficient use of this energy source (Chandarasekharam et al., 2014; Clark 2015). The technology is already cost-competitive in the U.S. with traditional sources of energy absent subsidies (Chandarasekharam et al., 2014); as such, it likely would be advantageous for regional governments to invest in it at a rate sufficient to spur substantial growth.

Biotechnology represents another potentially significant growth area for the GCC, and certain governments in the region – including Qatar through its Sidra Medical and Research Center and Science and Technology Park, Dubai through its Biotechnology and Research Park, and Abu Dhabi through the Institute of Science and Technology as well as its venture capital fund in Masdar – have begun to invest in this sector, although some has been dedicated to attracting international firms rather than promoting novel research (Birch, 2009; Dubai Biotechnology and Research Park, 2015). Given the long R&D cycles in this sector, it is difficult to assess the current effect of the investments.

Nevertheless, biotechnology holds substantial promise to enable the generation of novel intellectual property in fields such as genomics, bioinformatics, biopharmaceuticals, and agricultural biotechnology. The rapidly accelerating decreases in the costs of human genome sequencing resulting from rapid improvements in both sequencing machines and database technology make biotechnology particularly promising given its commercial applications in healthcare, agriculture, and energy. The high-risk nature of individual initiatives in this sector can be mitigated through a portfolio approach, as well as through strengthening domestic regulatory frameworks and intellectual property rights.

Nanotechnology constitutes another promising area of research for GCC countries. It overlaps significantly with other key prospective technological areas for the region, and as such, would be a complementary area for inclusion in educational curricula, scientific research parks, and venture capital portfolios. Within renewable energy, nanotechnology could be used to increase the strength of drilling machines for geothermal energy, or the strength or lightness of blades and photovoltaic panels for wind energy (Hussein, 2015). Within biotechnology, nanobiotechnology can be used for drug delivery, diagnostics, improved imagery, biosensing, and tissue engineering (Koopmans and Aggeli, 2010).

Nanotechnology also holds significant promise for improving information and communication technologies throughout the region. Several Gulf states have initiated steps to promote the space, including Saudi Arabia through centers at King Fahd University of Petroleum and Minerals, King Abdul Aziz University, King Abdullah University of Science and Technology, and King Saud University; Kuwait at its Institute for Scientific Research; Qatar through its foundation; and the UAE at the Masdar Institute and the Khalifa University of Science, Technology and Research (Alfeeli et al., 2012). Nevertheless, there remains significant opportunity for regional growth in research in the field.

Burgeoning efforts to promote certain of these sectors have attracted significant attention, but at present, as noted above, few results, whether measured by technical criteria such as patents or scientific publications, or commercial criteria such as revenues, profits, startups founded, or IPOs.

Conclusion/Policy Recommendations

While GCC governments recognize the importance of economic diversification and have begun to take steps to promote sustainability and advanced technology, they largely have done so through controlled, top-down approaches, rather than taking a systemic, structural perspective to addressing the challenges of sustainable development.

A first key measure would be to reduce dramatically the annual fossil fuel subsidies, which comprise a vastly greater portion of regional GDP than R&D spending, either public or private. The recent sharp downturn in energy prices presents a unique opportunity to do so even further. Rather than directly allocating the capital to subsidizing an immediate social service, the governments should allocate it to long-term technology development targeted to generating solutions to regional energy, environmental, and healthcare challenges.

A second measure would be to construct labor policies that favor the inward migration of talented expatriate scientists and engineers, and to recruit them actively (Smith, 2015). Although the Gulf states have a high proportion of international workers, they primarily consist of financial and corporate management professionals, on the one end, and migrant laborers on temporary visas, on the other. The creation of a vibrant scientific and technological ecosystem would itself help address the converse problem, of providing incentives for highly skilled native scientists and engineers to remain within the region; the government also ought to promote educational exchanges at the graduate and post-graduate levels to encourage knowledge transfer (Smith, 2015). Discussions of the refugee crisis in presently war-torn Arab countries have viewed it largely as a potential challenge for other regional governments to confront, rather than as agents who can make valuable contributions to other societies. Regional governments ought to consider taking in scientists and engineers from countries such as Syria, Iraq, Libya, and Yemen for public policy and economic as well as humanitarian reasons.

A third measure would be to engineer legal reforms that favor strong intellectual property rights, swift bankruptcy protection measures, and modern contract enforcement procedures, as well as measures to promote crowdfunding, angel investing, and startup incubation and tax incentives for entrepreneurship and scientific research and development.

A fourth would be to legally enshrine limits on annual resource extraction, to be phased in gradually over a 5 year period, that would help reduce societal dependence on the sector, promote conservation, increase fiscal stability, and reduce negative externalities associated with the sector, including pollution and climate change.

A fifth measure would be to develop advanced data systems that facilitate real-time tracking of sustainability measures and the development of technology-related startups, to enable precise and target interventions that promote sustainable growth.

A sixth would be to promote the development of infrastructure that is specifically targeted to technology entrepreneurship across multiple sectors, including advanced communication networks, research facilities and laboratories, incubators, accelerators, and graduate and post-graduate universities.

While the resource endowments of the GCC countries enable a high current standard of living, there is an increasing recognition that economies based principally on resource extraction are neither economically nor environmentally sustainable. The promotion of knowledge-based economies would have the additional benefit of fostering an increasingly knowledge-based society, in which evidence and data are accorded the highest authority in policy-making by government officials and citizens alike. The promotion of technology entrepreneurship through the reallocation of energy subsidies to R&D, labor, legal, and regulatory reforms, advanced data systems, and technology-related infrastructure can play a critical role in the promotion of sustainable development in the Gulf.

References

Ahmed, A., and Alfaki, I. (2013) Transforming the United Arab Emirates into a Knowledge-Based Economy: The Role of Science, Technology and Innovation. *World Journal of Science, Technology and Sustainable Development* 10(2): 84-102.

Alfeeli, B., Mohiuddin, T., and Saoud, K. (2012) Current Status of Nanotechnology in Arab Gulf States. *12th IEEE International Conference on Nanotechnology* August 20 Accessed at: http://tinyurl.com/hctedpp

Al-Roubaie, A. and Al-Zayer, J. (2006) Sustainable Development in the GCC Countries: the Impact of Technology Transfer. *World Review of Entrepreneurship, Management, and Sustainable Development* 2(3): 175-88.Al-Shuaibi K (2015) Ministry of Oman to the United Nations, Private Correspondence.

Arvai, Z., Prasad, A., and Katayama, K. (2014) *Macroprudential Policy in the GCC Countries.* IMF Staff Discussion Note.

Ben-Rubeian, A. (2015) Saudi Arabia's Falling Oil Revenues *Al-Monitor*, March 2. Accessed at: http://tinyurl.com/horb9mp

Beutel, J. (2012) Conceptual Problems of Measuring Economic Diversification, as Applied to the GCC Economies in Giacomo Luciani, ed. *Resources Blessed: Diversification and the Gulf Development Model.* London: Gerlach Press.

Birch, H. (2009) Gulf Beckons Biotech *Nature Biotechnology* 27.10: 879.

Biswas, R. (2004) Making a Technopolis in Hyderabad, India: The Role of Government IT Policy *Technological Forecasting and Social Change.* Vol. 71(8): 823-35.

Brinkley, I., Hutton, W., et al. (2012) *Kuwait and the Knowledge Economy.* The Work Foundation and the Kuwait Program on Development, Governance, and Globalization in the Gulf States. 22. The London School of Economics and Political Science. Accessed at: http://tinyurl.com/ycxue53v

Callen, T., Cherif, R., et al. (2014) *Economic Diversification in the GCC: The Past, the Present, and the Future.* Institute for Capacity Development and Middle East and Central Asia Department. International Monetary Fund. Staff discussion note. Accessed at: http://tinyurl.com/y9vpgfjc

Chandarasekharam, D., Lashin, A. and Al-Arifi, N. (2014) CO_2 Mitigation Strategy Through Geothermal Energy, Saudi Arabia. *Renewable and Sustainable Energy Reviews* 38: 154-163.

Charles, C., Moerenhout, T. and Bridle, R. (2014) *The Context of Fossil Fuel Subsidies in the GCC Region and Their Impact on Renewable Energy Development.* Manitoba: International Institute for Sustainable Development.

Chen, K. and Kenney, M. (2007) Universities/Research Institutes and Regional Innovation Systems: The Cases of Beijing and Shenzhen. *World Development* 35(6): 1056-1074.

Cherif, R. and Hasanov, F. (2014). *Soaring of the Gulf Falcons: Diversification in the GCC Oil Exporters in Seven Propositions.* IMF Working Paper 14/177. Accessed at: http://www.imf.org/external/pubs/ft/wp/2014/wp14177.pdf

Clark, P. (2015) Kingdom Built on Oil Foresees Fossil Fuel Phase Out This Century *Financial Times.* Accessed at: http://tinyurl.com/ha5u46v

Coady, D., Parry, I. et al. (2015) *How Large are Global Energy Subsidies?* IMF Working Paper Accessed at: http://www.imf.org/external/pubs/ft/wp/2015/wp15105.pdf

Doh, S. and Kim, B. (2014) Government Support for SME Innovations in the Regional Industries: The Case of Government Financial Support Program in South Korea. *Research Policy* 43: 1557-69.

Dubai Biotechnology and Research Park Accessed at: http://www.dubiotech.ae/

El-Katiri, L. and Husain, M. (2014) *Prospects for Renewable Energy in GCC States: Opportunities and the Need for Reform.* Oxford Institute for Energy Studies: 1. Accessed at: http://www.oxfordenergy.org/wpcms/wp-content/uploads/2014/09/MEP-10.pdf

Ennis C. (2015) Between Trend and Necessity: Top-Down Entrepreneurship Promotion in Oman and Qatar *The Muslim World* 105(1): 116-138.

From Regional Pioneer to Global Contender: The Economic Vision 2030 for Bahrain. Accessed at: http://issuu.com/economicdevelopmentboard/docs/bahrain_vision_2030

Gatete, C. (2015) *Exiting Fragility: Views From the Ground.* Speech at World Bank/IMF Spring Meetings.

The GCC in 2020: Resources for the Future (2010). New York: Economist Intelligence Unit. Accessed at: http://graphics.eiu.com/upload/eb/GCC_in_2020_Resources_WEB.pdf

Gelvanovska, N., Rogy, M. and Rossotto, C. (2012) *Broadband Networks in the Middle East and North Africa: Accelerating High-Speed Internet Access.* Washington, DC: The World Bank.

Hall, J., Daneke, G. and Lenox, M. (2010) Sustainable Development and Entrepreneurship: Past Contributions and Future Directions. *Journal of Business Venturing* 25: 439-48.

Harvie, C. and Lee, H. (2003) Export-Led Industrialisation and Growth: Korea's Economic Miracle, 1962-89. *Australian Economic History Review* 43(3).

Haug, M. (2014) Renewable Energy in GCC Countries: the Next Frontier? *Forum*, Oxford Institute for Energy Studies: 96.

Hausmann, R., Hidalgo, C. et al. *The Atlas of Economic Complexity: Mapping Paths to Prosperity.* Accessed at: http://atlas.cid.harvard.edu/

Helm, D. (2014) Taking Natural Capital Seriously. *Oxford Review of Economic Policy* 30(1): 109-25.

Hussein, A. (2015) Applications of Nanotechnology in Renewable Energies – A Comprehensive Overview and Understanding. *Renewable and Sustainable Energy Reviews* 42: 460-476.

Hvidt, M. (2011) Economic and Institutional Reforms in the Arab Gulf Countries. *Middle East Journal* 65(1). Accessed at: http://www.jstor.org/stable/23012095

Hvidt, M. (2013) Economic Diversification in the GCC Countries. *London School of Economics* 27.

Hvidt, M. (2015) The State and the Knowledge Economy in the Gulf: Structural and Motivational Challenges. *The Muslim World* 105(1): 24-45.

Kagame, P. (2010) The Backbone of a New Rwanda: Entrepreneurship is the Surest Way. *Innovations: Technology, Governance, Globalization* 5(1) 3.

Kern, S. (2012) GCC Financial Markets: Long-Term Prospects for Finance in the Gulf Region. *Deutsche Bank Research.* Accessed at: http://tinyurl.com/jgs59yp

Kerr, S. and Clark, P. (2015) UAE drops fossil fuel subsidies to boost finance and curb omissions. *FT.* Accessed at: http://tinyurl.com/gpmsnrq

Koopmans, R. and Aggeli, A. (2010) Nanobiotechnology – quo vadis? *Current Opinion in Microbiology* 13(3): 327-334.

Lagarde, C. (2015) A Conversation on the Global Economy. *World Bank/IMF Annual Meetings*.
Mueller, H. (2012) GCC Industrial Development in Giacomo Luciani, ed. *Resources Blessed: Diversification and the Gulf Development Model*. London: Gerlach Press.
Murenzi, R. and Hughes, M. (2006) Building a Prosperous Global Knowledge Economy in Africa: Rwanda as a Case Study. *Int. J. Technology and Globalisation* 2(3/4): 252-265.
Norman, M. (2014) Have Special Economic Zones Entered the 21st Century Yet? A Tale of Two Cities. *World Bank Private Sector Development Blogs*. Accessed at: http://tinyurl.com/nkncodn
Osman, D. (2015) The State and Innovation: An Analytical Framework. *The Muslim World* 105(1).
Pereira, A. (2008) Manufacturing Human Resources. *Social Policy in Post-Industrial Singapore* Boston: Brill 121-44. Accessed at: http://tinyurl.com/gthoq2l
Rwandan Economic Development Board (2014) *Investing in Rwanda: an Overview*. African Development Bank Annual Meeting. Accessed at: http://tinyurl.com/jk8xab3
Sale, P., Feary, D. et al. (2011) The Growing Need for Sustainable Ecological Management of Marine Communities of the Persian Gulf. *Ambio* 40(1): 4-17. Accessed at: http://www.jstor.org/stable/41417242
Sanchez, J., Salinas, A. and Harris, J. (2011) Education with ICT in South Korea and Chile. *International Journal of Educational Development* 31: 126-48.
Saxenian, A. (2006) *The New Argonauts*. Cambridge: Harvard University Press.
Schwab, K. and Sala-I-Martin, X. (2015) *The World Economic Forum Global Competitiveness Report 2014-5*. Accessed at: http://www3.weforum.org/docs/WEF_GlobalCompetitivenessReport_2014-15.pdf
Sharif, N. and Tang, H. (2014) New Trends in Innovation Strategies at Chinese Universities in Hong Kong and Shenzhen. *Int. J. Technology Management* 65(1-4): 300-318.
Shenzhen Government Online *Four Pillar Industries in Shenzhen*. Accessed at: http://english.sz.gov.cn/iis/201408/t20140815_2546625.htm
Sillitoe, P. ed. (2014) *Sustainable Development: An Appraisal from the Gulf Region*. New York: Berghhahn Books.
Smith, M., Executive Director, Iraqi Women's Fellowship Foundation (2015) Private Correspondence.
(2015) South Korea *Nature* 519 (7544), S62-S63. Accessed at http://www.nature.com/nature/journal/v519/n7544_supp_ni/full/519S62a.html
UAE Vision 2021 Accessed at: http://www.vision2021.ae/en
UNDP Arab Knowledge Report 2014: Youth and Localization of Knowledge (2014) Dubai: Al Ghurair.
World Bank Data (2015) *CO_2 Emissions per Capita 2010*. Accessed at: http://tinyurl.com/kte54ku
World Bank Data *Market Capitalization of Listed Companies*. Accessed at: http://tinyurl.com/jggx2f3
World Bank Data. *Adjusted Savings: Natural Resource Depletion (% of GNI)*. Accessed at: http://tinyurl.com/h67jwqs
World Bank Doing Business 2015. Accessed at: http://www.doingbusiness.org/rankings
Zweig, D. (1997) To Return or Not to Return. *Studies in Comparative International Development* 32(1): 92-125.

3

The Development of Knowledge-Based Centers in the GCC

Pia-Kristina Anderson

Introduction

In the last few decades the states of the Gulf Cooperation Council (GCC) have enjoyed economic success as both oil producers and trade hubs. Aware that their main advantage in the present is a non-renewable resource base that is currently providing wealth but will not continue to do so indefinitely, these states are looking to the realm of knowledge as a base for their future economies. In recent years several of the GCC states, notably the small but wealthy states of UAE and Qatar, have made efforts to use their locations, relatively stable political situations and natural resources to develop their human capital by investing in a variety of ambitious projects.

While the GCC states have invested their money widely and diversely, they have made significant investments in education, scientific research, and heritage. The reasons for this are varied: more enduring sources of wealth and prestige are undoubtedly part of the motivation, but so are concerns about the continuation of their culture, the reliance on foreign workers to fill many private sector jobs, and the stability of the wider Middle Eastern region. Alongside these pragmatic concerns are more idealistic motivations, such as a genuine desire for more intellectual and cultural pursuits, a love of their heritage and even an attempt to bring about a new golden age of Arab-Islamic science and knowledge. Additionally, the existential threat posed by regional and global disorder is an under-examined contributing factor for the rapid development of knowledge centers at considerable expense; these centers offer a positive alternative model of Arab society. Countries such as the UAE and Qatar seek to set themselves up as future exporters of knowledge as a way of protecting their prosperity, and thus their stability.

In the last few decades numerous and varied centers of knowledge have been built. Examples include Universities, Science Centers and Renewable Energy Research Centers. Healthcare centers such as Dubai's Heathcare City have set up research arms such as the Mohammed Bin Rashid Academic Medical Center to focus on the specific health needs of the region.

To develop these knowledge centers, land grants (frequently with a special 'free zone' status) are given to themed 'cities' that house these centers: Abu Dhabi's Masdar City, Dubai's Knowledge Village and Sharjah's University City being examples. Capital is often provided to develop not only infrastructure, but also to offer high salaries to entice international scholars on short or long-term contracts, and to provide scholarships for local students and some promising expatriates. A theme of the past few decades in both universities and corporations is that scientists and scholars are brought from the West (or trained at Western institutions) while laboratory and middle-range expertise is often brought from the Indian subcontinent, the larger Middle Eastern region and the Philippines. This theme is being challenged by the new generation of GCC nationals that are being trained to take over positions in higher education.

The development of these knowledge-based centers has to be 'sold' to various parties; investors, researchers and academics, the media or the general public. Consequently the initiatives themselves are accompanied by a great deal of publicity in both English and Arabic aimed at different groups.

While a main theme of the burgeoning knowledge centers is to put the region on the global stage, several of the centers have national culture and identity as their focus. Among these are arts and heritage organizations that have concentrated on the long-term traditions and cultures of the relatively new states. While global art is being set forth in some of the new museums, also emphasized are heritage institutes, Islamic museums, and Arabic language publications.

Book fairs and literary centers have also become a much-anticipated part of GCC culture for both nationals and expatriates, often filling in for the regional scarcity of bookstores. Arabic literacy has become an issue in countries where English is widely spoken. The demand for literary events and good Arabic books has attracted attention from the private sector, charities and government bodies.

All of these varied developments take place in an environment of patronage and competition that has some historical parallels. One point of comparison is the Italian Renaissance. Like the modern GCC, fifteenth century Italy comprised several states that shared a common language and culture but still competed for supremacy with achievements in art, architecture and science demonstrating the prestige of the different city-states and rulers. The parallel is not exact – the GCC is a formal political union in which the members work towards common goals, and competition between the states stays in the realm of which country attains the greatest level of prestige and international reputation. Nevertheless, similar motivations to those which drove the Renaissance princes are comparable in the GCC today.

The Renaissance is a useful lens for examining the competitive aspect of developments in the GCC countries and the role of leaders as patrons of the arts and sciences. However, when discussing the nation-building projects that these states are undertaking, there is another parallel that the leaders of the GCC states themselves are quick to call upon when announcing their ambitious goals: the Islamic Golden Age, when the Islamic caliphate drove innovation in arts and sciences. The modern GCC states are keen to make comparisons to the Golden Age, and it is invoked frequently in the grand scientific projects that have been announced in recent years, such as the UAE's space program.

This chapter will examine a number of the initiatives coming out of the GCC states, particularly the UAE, through this lens of political patronage. The UAE makes a particularly interesting case study as, along with its wealth, it has an unusual government in the form of its federal monarchy, with each of the seven Emirates ruled by an emir and his family, and each emir lending some of his character to his Emirate. Furthermore, the fact that the rulers of the Emirates come from formerly rival tribes who only since 1971 began working together has led to the creation of a large-scale national identity-building project, with an important milestone imminent in the form of the UAE's 50th anniversary in 2021, meaning that many of the UAE's aspirations will soon be tested as their completion date nears. This looming date has shaped two initiatives in particular: Dubai's hosting of the World Expo in 2020, and the Emirates Mars Mission, which will launch the 'Hope' probe in 2020 for a 2021 arrival in Martian orbit. These events are shaping a great deal of other activities in the UAE's pursuit of a knowledge-based society.

Universities

A recurring theme of knowledge initiatives in the GCC is that the rulers aspire for their countries' achievements to be beacons to the rest of the Muslim world. However, they start from a disadvantageous position. The Zakri Task Force on Science at Universities in the Muslim World examined the countries with a Muslim majority that make up the Organization of Islamic Cooperation (OIC). The study of the 57 OIC countries shows that at present these countries are lagging far behind the international average in investment in scientific research, with less than 0.5% of their GDP invested in R&D compared to the average of 1.78% (Muslim World Science Initiative, 2015: 27). According to the 2014/2015 QS World University Rankings, no OIC university is in the top 100. Of the 17 OIC universities in the top 400, only 5 are in GCC countries – King Fahd University of Petroleum and Minerals, King Saud University and King Abdulaziz University in Saudi Arabia, and United Arab Emirates University and the American University of Sharjah in the UAE (Muslim World Science Initiative, 2015: 26). No GCC universities appear in the Times Higher Education top 400, which uses different criteria that emphasize research (Muslim World Science Initiative, 2015: 26-27). There is a context to these low scores by GCC states: the Task Force reported that "Three quarters of all Arab universities were established in the last 25 years of the 20th century. Fifty-seven percent of them are no more than 15 years old" (Muslim World Science Initiative, 2015: 37-38).

Conscious of this lag behind the rest of the world, the GCC states have set about making major investments in higher education to rectify the deficiency. Countries like the United Arab Emirates (UAE) and Qatar have used petrodollars, trade-oriented economies and their strategic geographic positions to fund and found universities and research centers, bringing in transnational expertise both by hiring international faculty and by partnering with foreign institutions. One of the main objectives in the creation of universities is the education of national citizenry, providing them with skills and knowledge so that they may be gainfully employed in their own countries. Initiatives such as 'Emiratization', 'Qatarization', and 'Saudization' aim to ensure that educated nationals are able to succeed in professional, research and leadership positions in their own countries, with the expectation that eventually educated locals could hold many of the specialized positions currently held by expatriates.

The GCC universities emphasize educating the local population, but also cater to the large expatriate populations in the region. While universities throughout the GCC vary greatly in specialization, educational model and quality, together they have made the GCC a draw for Muslim students worldwide wishing to study at a university with an international reputation, but in an Islamic country. The GCC is becoming a magnet for students from the larger Middle East region, South Asia, Central Asia, and North and West Africa looking for quality tertiary education.

GCC-based higher educational institutions also offer a space for GCC women to obtain undergraduate and graduate education in or near their home countries, thereby bypassing social and cultural pressure that might hinder them from studying internationally.

The universities in the GCC represent a mixture of instructive and educational styles. One of the most successful models for undergraduate institutions is the American liberal arts form of education. This form of 'American University' education aims to provide students with a broad general education and lessons in life-long learning techniques while providing a deep knowledge of the student's chosen major field of study. The American liberal arts education also proposes to offer the students an education in social responsibility and critical thinking, while providing skills such as analytical problem solving and a broad range of communication skills. Additionally, American style universities provide 'student life' activities and are usually co-educational.

Most American universities in the GCC prefer faculty who are trained and/or have their doctorate or final degree from an American institution, which tends to necessitate a transnational faculty from a variety of countries, but with North American degrees. The American University of Sharjah is an example of such an institution.

British and Australian universities have also been established in the GCC. These styles of education tend to focus the student more towards their major field of study, but also aim to provide a grounding in writing and analytical skills. Middlesex University Dubai is an example of a British style university, whereas the University of Wollongong in Dubai offers an Australian style of education. Some universities, such as the British University in Dubai, specialize in postgraduate (or graduate) education and offer specialized programs associated with selected universities in the UK in the fields of Business, Education and Engineering (Buid.ac.ae, 2015).

European Universities are also represented in the GCC; the Université Paris-Sorbonne Abou Dhabi provides a French university system in both French and English languages (Sorbonne.ae, 2015).

Beyond differing national styles of education, a chief variance is whether the university is home-based or a branch campus. Examples of home-based universities are the American University of Sharjah, UAE; King Saud University in Riyadh and King Abdullah University of Science and Technology in Thuwal, both in Saudi Arabia; and the American University of Kuwait. Branch campuses include New York University Abu Dhabi, Texas A&M University at Qatar and Weill Cornell Medical College in Qatar. Home-based universities vary with branch campuses specifically in that branch campuses reflect the character and curricula of their parent institutions, while home-based universities are independent and not administratively linked to US or other universities. Institutional focus on a particular field, such as engineering, science, business and management or the arts, is often imposed by either the parent university or the local board of trustees or founder of the home university.

There is also the distinction between private and public universities. Public institutions are sponsored by local governments and tend to share characteristics such as a largely, if not exclusively national student body, and gender segregation. Examples of public institutions are United Arab Emirates University, Zayed University, and the Higher Colleges of Technology in the UAE, Qatar University, and other similar ones in the region. Meanwhile private institutions draw from many nationalities and usually are co-educational.

An important distinction amongst universities is whether the institution is for profit or a non-profit institution. In the case of the for profit institutions, the type of educational system, the nature of the faculty, the fields of study, and the range of students accepted to the university is influenced by the amount of projected profit for the institution. Non-profit universities rely more heavily on scholarship students and endowments. Further, many Gulf universities are encouraged to focus on specific research and development themes by patrons who guide research in fields such as architecture and design, health or renewable energy. This connects the educational systems of the Gulf nations with the political and commercial sector of the GCC.

One of the main connections of GCC Universities to other international universities is through MOUs, which often have specific provisions for the exchange of scholars and graduate students, and for the use of facilities such as a Gulf base for international conferences. Universities also work with Gulf-based think-tanks to develop research platforms to host conferences, and work with international corporations to establish Gulf centers for research and development.

Science Centers

The Emirates Mars Mission is perhaps the most explicit statement of the UAE's intent to turn its wealth into national scientific expertise. The UAE tends to anchor its long-term ambitions on 'mega-projects', and the UAE Mars Mission is perhaps the most ambitious of these yet – a country with little history of space travel sending a probe to Mars by 2021. This grand gesture is however merely a part of a twofold ambition to develop the UAE's presence in space, and its overall presence in the sciences. The former is demonstrated by the already-orbiting satellites launched in recent years and the forthcoming KHALIFA SAT imaging satellite due for launch in 2018, while the latter is stated explicitly in the organization's desire to "leave behind a valuable and enduring legacy in the form of human capital: a generation of experienced scientists and engineers trained and inspired by the Mars mission" (Emiratesmarsmission.ae, 2015).

The Mars Mission is very much a case of the UAE investing its substantial financial capital in importing scientific expertise and partnering with prestigious foreign institutions, in order to develop its national scientific capacity. While engineering is a popular subject at Gulf universities, the region is not known for valuing or excelling in the sciences. The UAE is seeking to change this in a generation using the glamor of space exploration to entice young Emiratis into studying science and pursuing it as a career. To this end it is also highlighting the fact that while the 2021 mission will inevitably rely on outside expertise, it will be led by a team of Emiratis who will also be the public face of the mission. Building scientific capability from scratch requires a substantial element of public relations work alongside the research and development of the mission itself. This can be seen not only in the highlighting of the existing Emirati team (set to expand substantially over the years to a team of 150 engineers and scientists), but also in the emphasis on how the mission offers hope, wealth, and a chance for the whole Arab people to be "contributors to humanity and civilization" (Emiratesmarsmission.ae, 2015).

Even the date of the mission is symbolic – 2021 will be the 50th anniversary of the United Arab Emirates, while the launch of the probe will take place in 2020. While the launch window in 2020 is dictated by the pragmatic concern of when the distance between Earth and Mars is at its smallest, it is also the year that Dubai hosts the World Expo and a date that has in recent years become significant to UAE, both for commercial reasons and as part of the UAE's national mythology – a source of pride in accomplishments leading to the successful bid and the promise of transformative future progress. This presentation of the UAE as a young and dynamic country moving forward feeds back into its attempts to recruit and train a generation of scientists. In the 2014 press release announcing the mission, the chairman of the Defence Services Marketing Council was quoted as saying "The decision will lead the nation into a visionary period for decades ahead that will encourage more Emiratis to become part of the global solution that the entire planet is searching for beyond Earth" (Emiratesmarsmission.ae, 2015a).

The actual nature of the mission is noteworthy. Beneath the use of large numbers promising dramatic results – "1000 Gigabytes of new data about Mars shared over two years" (Emiratesmarsmission.ae, 2015b) – is the more concrete and immediate promise of the design and construction of the probe itself, and the details of the mission's scientific goals: a study of Mars's atmosphere, an investigation of why it is losing its atmosphere to space, and the creation of "the first global picture of how the Martian atmosphere changes throughout the day and between the seasons" (Emiratesmarsmission.ae, 2015b). This development of expertise and accumulation of knowledge is being offered to the world for future collaboration – the UAE's Vice President Sheikh Mohammed bin Rashid al Maktoum has framed the mission in terms of "how positive a contribution the Arab people can make to humanity through great achievements, given the right circumstances and ingredients" (Emiratesmarsmission.ae, 2015a).

This interest in space sciences is visible elsewhere in the UAE, notably in the recent opening of the Sharjah Centre for Astronomy and Space Sciences (SCASS) in Sharjah University City. Comprising a planetarium, an observatory, and astronomy and space exhibitions, alongside the more unusual features of "Exhibition of the Universe in the Holy Qur'an … and a Cosmic Park", (SCASS, 2015), the center serves both the general public and researchers at the Sharjah universities. It fits very much with Sharjah's strategy of using museums and heritage both to attract tourists and to enrich the lives of the public – evoking Islam's golden age of scientific research but also reviving it through activities aimed at students and children. Professor Hamid Al Naimiy, chancellor of the University of Sharjah, has stated as much: "Our goal is to inspire Arabs to dream again, to venture into the world of astronomy and space sciences", and "It is not enough to say we had a glorious, golden Islamic civilization of innovations. That was so long ago; we need to do something now and for the future"(Ghazal, 2015). As with the Mars Mission, there is a distinct thread of reviving something historic that runs through the establishment of UAE scientific projects.

The UAE is not alone in pursuing science and technology. Qatar's Science and Technology Park (QSTP) – complete with catchy slogan "where research goes to work" (Qstp.org.qa, 2015) – seeks to "be a recognized international hub for applied research, innovation and entrepreneurship" (Qstp.org.qa, 2015a). QSTP "aims to grow Qatar's 'post-carbon economy'" and "focuses on four themes … Energy, Environment, Health Sciences and Information & Communication Technologies" (Qstp.org.qa, 2015). The QSTP presents itself as a coming-together of research, education and business – with the promise of wealth creation very much focused on attracting international investment.

As with the UAE's scientific organizations, marketing plays an extensive role in selling QSTP to investors and the public at large. It is important to note that this marketing does not simply serve as corporate gloss – the Gulf countries are late to enter the science and technology field and marketing serves the important purpose of highlighting the advances made by these countries and counteracting entrenched stereotypes about the Middle East. QSTP's "trailer", available online, is an example of a common approach of the Gulf's to large projects – a juxtaposition of futuristic architecture, Arabs in traditional Gulf dress alongside Western businesspeople in suits, and business-oriented language such as "competitive advantage", "hub". and "entrepreneurship" (QSTP, 2013). Traditional marketing is not the only approach they use to raise their profile – a series of "TECHtalks" on "trending" issues are held, free of charge, to draw attention to the areas of research currently being pursued which are of particular interest to GCC citizens and residents, such as how genetic research can help fight obesity and diabetes, prevalent in the region (QSTP, 2015).

Likewise, comments by Qatar's Sheikha Moza bint Nasser Al Missned recall the Islamic golden age as a model for Qatar's role in the world: "How can we build the future, how can we be active players in the global map? It is not a changing of culture, it is a resurrecting of old culture, of a very positive culture that we used to have" (Qatar Foundation, 2012). Qatar here presents its oil and gas resources as a blessing which can be reinvested in society – in education, and in creation of a place where global organizations can gather and find solutions to the world's problems (Qatar Foundation, 2012). While this could be seen as an extension of Qatar's general policy of using oil wealth to punch above its weight in arenas ranging from culture (Al Jazeera) to sport (the FIFA World Cup), investing in science and technology gives Qatar additional moral weight for looking to the future and offering to share its wealth with the world.

Abu Dhabi's Masdar City has a very particular focus – renewable energy. Abu Dhabi is rich in oil, but also in desert, and Masdar is an attempt to take advantage of year-round sunshine to build a sustainable city – carbon-neutral and drawing its energy from a 22-hectare field of solar panels (Kingsley, 2013). While the financial crisis has slowed Masdar's development as a full-scale city, Masdar has been successful in developing the Masdar Institute for Science and Technology (MIST), a postgraduate university modelled on the US graduate system and working in cooperation with the Massachusetts Institute of Technology (Masdar.ae, 2015; Kingsley, 2013). Masdar co-founder Steve Geiger has commented on how MIST is "already graduating first-class students doing pretty near first-class research" (Kingsley, 2013), while MIST itself reports that since 2009 "3 patents have been approved so far, while 44 patent applications are pending and 68 invention disclosures have been received" (Masdar.ac.ae, 2015). In 2014 MIST's student body comprised "491 students from 66 countries", with 44% of students being UAE nationals and nearly 44% of the students being female (Masdar.ac.ae, 2015). The areas of research are driven by Masdar's priorities, and in some cases backed by corporate or government bodies (Masdar.ac.ae, 2015a). While it is too early to tell what Masdar's long-term prospects are, its success in establishing a reputable postgraduate institute focused on STEM subjects and attracting a diverse student body is highly encouraging.

As with the Mars Mission and QSTP, Masdar City pointedly calls back to an older tradition – linking cutting-edge science and technology with an older (and, implicitly, wiser and more decent) time, rather than simply presenting it as the next step in the UAE's rapid modernization. *Wired* magazine in a 2013 article highlights this ancient-modern synergy: "a 45-metre-high wind tower, modelled on age-old Middle Eastern designs, sucks air from above and pushes a cooling breeze through Masdar's half-dozen streets ... the buildings are situated close together, unlike many developments in the UAE, creating shaded, pedestrianised thoroughfares" (Kingsley, 2013). And despite the difficulties in establishing a full-scale city on these principles, it has made something of a difference to the region's discussion of sustainability – the World Wildlife Fund welcomed the initial, highly ambitious announcement of Masdar City (WWF, 2008). Its lasting impact has been to encourage other Arab countries to take sustainability seriously, "because it's not only environmental, it's also economic" (Kingsley, 2013). Likewise, Masdar has served as successful international marketing for the UAE's technological prowess – its self-driving cars in particular are included in tours for visitors and were featured in *Brave New World with Steven Hawking* as a vision of the future (Hawking, 2011), and even play a part in the UAE's diplomatic efforts, with Indian Prime Minister Narendra Modi riding in one as part of a state visit (Angre, 2015).

While the GCC states are competing with each other – in the case of the UAE, there is even the well-known competition between Abu Dhabi and Dubai, a competitive ambition that could be said to drive innovation – they perhaps do not get enough credit for such internationally collaborative efforts as the scientific and technological ventures detailed here. It is to be hoped that cooperation between Gulf institutions and the wider world can continue increasing.

Arts and Heritage Organizations

Although technology is a main focus, the GCC States are also interested in building creative knowledge centers for the arts and heritage fields. Museums have traditionally served as "store houses of knowledge" (Cannon-Brookes, 1984), storing and displaying objects in order to pass on knowledge and serve as research collections for the creation and dissemination of further knowledge. In the GCC, they also play a substantial role is codifying and communicating the shared culture of these young countries, for the benefit of both citizens and tourists. The Emirate of Sharjah in particular is notable for making cultural tourism a key part of its museums.

Sharjah's numerous museums have been effectively educating and entertaining both the general public and school children for decades, while simultaneously serving as research, conservation and heritage centers. This relatively long tradition by the Emirate of Sharjah on culture in general, traditional Emirati heritage in particular, and on both Arabic and Western educational systems, has led to this Emirate being selected by UNESCO in 1998 as Cultural Capital of the Arab World and as the Islamic Culture Capital in 2014 by the Islamic Educational, Scientific and Cultural Organization (ISESCO) (Sicc.ae, 2015). Sharjah is also a popular destination for Islamic tourism and emphasizes its family friendly and educational attractions. The Sharjah Museums focus on themes such as heritage – cultural, architectural, calligraphic, and archaeological; science – with both a science museum and discovery center; and religion – with the Sharjah Museum of Islamic Civilization being one of the largest and most visited of the museums in the UAE. Natural history is also preserved and studied at the Sharjah Aquarium, the Breeding Center for Arabian Wildlife, the Natural History Museum, the Botanical Museum and the Children's Farm. Finally the Arts are housed in the Sharjah Art Museum, which also sponsors an arts center for local and international artists. The Sharjah Art Foundation has become recognized internationally as providing a base for art practitioners in the region. It provides a steady series of public events as well as a venue for art publication (Sharjahart.org, 2015).

The main focus of Abu Dhabi's cultural promotion is the massive infrastructural development of Saadiyat Island off the Abu Dhabi coast. The centerpiece of Saadiyat Island is the Zayed National Museum, dedicated to Sheikh Zayed bin Sultan Al Nahyan, the founder of the United Arab Emirates. Due to open in 2016, it is intended to serve as an educational and research center. At the time of writing its website prominently features a quotation from Sheikh Zayed: "The prosperity that we have witnessed has taught us to build our country with education and knowledge and nurture generations of educated men and women" (Zayednationalmuseum.ae, 2015). Also emphasized are "the long history of the region and its cultural connections across the world" (Zayednationalmuseum.ae, 2015a) – a significant comment when juxtaposed with the Foster + Partners-designed architecture of the museum, which is designed to resemble "the dynamic of flight and the feathers of a falcon" (Zayednationalmuseum.ae, 2015b). As with Masdar City and the other scientific projects, a larger Arabic history is invoked, giving legitimacy to ambition.

The Louvre Abu Dhabi opened its first major exhibit in April 2014 and is currently being established as a "universal museum" with exhibits from antiquity through to contemporary art and design (Louvreabudhabi.ae, 2015). The museum's website hails "an unprecedented alliance between the United Arab Emirates and France, through one of the highest level of cultural cooperation ever created between two sovereign countries" and draws connections between the Silk Route, the Enlightenment, and Abu Dhabi's role as a "crossroads for civilisations" (Louvreabudhabi.ae, 2015). As with Zayed Museum, the Louvre Abu Dhabi is still under construction and we are yet to see the result of the cultural cooperation taking place, but the promise of a museum that showcases the interlinked history of the world and cultures is one of great value, and one that seems the fulfilment of the ideal of "storehouses of knowledge".

The United Arab Emirates has also focused some energy on promoting and developing contemporary culture. Dubai in particular has put a great deal of energy into annual festivals of literature, film and music – events which attempt to cater for Emirati, expatriate and tourist audiences, and which try to support local artists while also bringing in high-profile international speakers.

The Emirates Literature Foundation, established in 2013, is an attempt by the Dubai government and the organizers of the Emirates Airline Festival of Literature (Emirates LitFest) to go a step further and create a non-profit organization dedicated to fostering local writing talent and encouraging reading in both Arabic and English. Describing itself as "a catalyst for writing and cultural exchange" (Emiratesliteraturefoundation.org, 2015), the Emirates Literature Foundation is responsible for the annual festival of literature, the Dubai International Writers' Centre (DIWC), and one-off events such as the 2014 Nabati Poetry Conference and a translation conference planned for early 2016. The DIWC, in operation since 2014, balances its roles as an "all-inclusive celebration of the written word" and "international hub for writers" through a year-round program of writing classes and public events (Diwc.ae, 2015).

As is frequently the case with arts organizations in the GCC, private sponsorship is critical to the operations of the Emirates Literature Foundation – in this case Emirates airline, one of Dubai and the UAE's most successful companies, is investing a share of its sponsorship budget in the development of the UAE's culture and tourism industries. In addition to Emirates airline and government bodies such as the Dubai Culture and Arts Authority, private companies and international organizations have supported a number of aspects of the foundation's activities. Some of these are private companies with an existing relationship to literature – educational and commercial UK publisher Oxford University Press sponsors a children's writing competition (Emirateslitfest.com, 2015a), while Costa Coffee, sponsors of the UK's Costa Book Awards, supports the DIWC and sponsors year-round events and children's workshops at Emirates LitFest (Costa.co.uk, 2015; Emirateslitfest.com, 2015b). Others are local or international charities – for example the Hamdan Bin Mohammed Heritage Center, an organization which promotes UAE heritage and folklore and which acts as "heritage strategic partner for Expo 2020", provides cultural activities for visiting authors and sponsors a poetry recital competition which "follows the strong oral tradition which exists in the Arab world" (WAM, 2015; Emirateslitfest.com, 2015c).

While Dubai has successfully, and within the GCC uniquely, established a literature festival on the model of events such as the Hay Festival; Abu Dhabi, Sharjah, Doha, Manama, Muscat and Riyadh all run annual book fairs, primarily industry-facing events that seek to rival longer-established Middle Eastern book fairs in Cairo and Beirut. The Abu Dhabi International Book Fair is interesting for its longstanding cooperation with Frankfurt Book Fair, the largest and best-known book fair in the world – an example of the GCC bringing in outside expertise to advise on a relatively new venture, while German publishers benefit from exposure to new and valuable Middle Eastern markets (Adbookfair.com, 2015a; Adbookfair.german-pavilion.com, 2015). Abu Dhabi Book Fair also partners with cultural organizations such as the British Council and the Goethe Institute to run cultural events and writing workshops, adding prestige and expertise from these countries and widening the range of public-facing events (Adbookfair.com, 2015a, 2015b).

The importance of book fairs as cultural events in the GCC should not be underestimated – Middle Eastern book fairs have an important role in public intellectual and educational life. While they are largely focused on business and the making of rights and translation deals, their customer-facing side is critical for people seeking to buy books and educational resources in Arabic. The book trade in the Arab world is substantially less developed than in the English-speaking world, Europe or India, and while these regions exhibit at the fairs, customers at Sharjah Book Fair for example can be seen with supermarket trollies filled with books. Cairo-based journalist Marcia Lynx Qualey attributes this aspect of the fairs to "distribution problems between Arab countries, and the lack of strong bookstores in some countries", adding that "Publishers lug their books from fair to fair, and hungry readers can come see what's available" (Qualey, 2014). Qualey has commented on how book fairs in the GCC have expanded and professionalized, and serve as an "important place for sharing titles and meeting authors" (Qualey, 2014). While international media tend to focus on the larger initiatives, such as Abu Dhabi's International Prize for Arab Fiction (in association with Man, sponsors of the UK's Booker Prize) and Zayed Book Award, the consumer-facing side of the book fairs arguably has the larger impact on GCC citizens.

Archives and Libraries

In 1968 Sheikh Zayed, the founder of the UAE, set up the Center for Documentation and Research. This center had the important goal of collecting, documenting and preserving material related to the history and heritage of the Arabian Peninsula, Abu Dhabi and the Emirates (Na.ae, 2015). This center has evolved into the National Archives of the UAE. The National Archives has continued the Center's original aims of collecting and archiving documents, visual records and audio archives related to the history and heritage of the UAE, and is "the largest documentation center in the Arabian Gulf Region" (Na.ae, 2015a). It is currently open to the public – from students to local and international scholars and provides a wealth of information on Emirati culture and history. The National Archives provides publications in the form of books as well as the bi-annual peer-reviewed journal *Liwa*, and in addition conducts corporate and community outreach, with one of its major aims being "to promote values of identity, belonging and loyalty in the hearts of younger generations" (Na.ae, 2015b). This focus on youth is a frequently stated concern in the GCC countries, stemming from the fear that younger generations, born since the economic boom brought by oil wealth, are growing up with no attachment to their cultural heritage or mother tongue. The UAE has been successful in documenting its recent past, but is preoccupied with ensuring that future generations value it.

Another initiative that aims to preserve local traditions is the Zayed Center for Heritage and History, based in Abu Dhabi. This center has initiated a large oral history project that is currently collecting folklore, traditional poetry and music.

Also in Abu Dhabi is the National Library, which aims to encourage Emiratis to access international knowledge while stressing the importance of reading and writing the Arabic language. Its varied approaches include creating accessible, Arabic-language information, translating international publications into Arabic and publishing Emirati writers (TCA Abu Dhabi, 2015).

As education as well as entertainment becomes more globalized, Gulf governments have identified the importance of stressing the maintenance of cultural and linguistic identity. While older generations took heritage for granted, the generation currently in Universities is inundated with global culture and has become more comfortable in English. Initiatives to emphasize Arabic have begun in primary and secondary schools in the UAE, and Universities are urged to require Arabic heritage courses (taught in English or Arabic).

It is in this context that the UAE declared 2016 to be the Year of Reading, jointly announced by Sheikh Khalifa bin Zayed Al Nahyan and Sheikh Mohammed bin Rashid Al Maktoum and explicitly acknowledging that this is part of the UAE's nation-building project as well as implicitly acknowledging the UAE's ambition to be a scientific and cultural first among equals in the GCC and the Arab world in general, and even to be a global player in these fields. For example, an Emirates News Agency press release notes:

> The Year of Reading initiative continues the approach and vision of the founding father, the late Sheikh Zayed bin Sultan Al Nahyan, which was based on the conviction that nation-building is not just a focus on urban and technological development, but is highly dependable on human development, which is the real cornerstone of success, and is the reason why the UAE has been able to transform itself in a relatively short period of time, becoming a beacon of progress and centre of culture in both the region and the world. (WAM, 2016)

It is significant that whereas the existing literary events mentioned in the previous section were independent of each other and shaped by the interests of their patrons, they are now being incorporated into a national strategy shaped by the UAE's global ambitions that also includes a planned library that will be the Arab world's largest (shaped like an open book), investments in publishing and audiobook production and support for Emirati authors (Green, 2016; East, 2016; Rostamani, 2016). Some of these initiatives – those involving book publication and distribution, for example – have quantifiable outcomes and it will be interesting to see if they are successful, and if they lead to a change in reading culture in the UAE.

Sheikh Mohammed Bin Rashid Al Maktoum Foundation

Dubai's Sheikh Mohammed Bin Rashid Al Maktoum Foundation (MBRF), set up by Sheikh Mohammed Bin Rashid Al Maktoum, vice president of the UAE and ruler of Dubai, is an example of a GCC ruler taking a direct hand in educational matters. Set up with a record-breaking $10bn donation (Leyne, 2007), the foundation is charged with closing the "wide knowledge gap" between the Arab world and the "developed world in the West and Asia" (Aljazeera.com, 2007). Using this money, the foundation "aims to empower future generations to devise sustainable solutions that address challenges in the knowledge and research domains in the Arab world" and is "committed to disseminating knowledge throughout the region through fostering ideas and innovation and funding research projects, activities and initiatives" (Mbrfoundation.ae, 2015a). The MBRF's initiatives include the Dubai International Writing Program, supporting 100 Emirati writers; My Family Reads, which distributes books to Emirati families; and the Bil Arabi Initiative, which encourages young Arabs to use Arabic rather than English on social media (Mbrfoundation.ae, 2015a); its most recent effort on this front being the "Dubai Digital Library", a platform for Arabic ebooks "including translations, as well as Arabic and foreign periodicals, dictionaries, biographies, and maps" (Mbrfoundation.ae, 2015b).

In addition to the above ventures, aimed primarily at Emiratis and other Arabic-speakers, the MBRF collaborates with the United Nations Development Programme (UNDP) on the Arab Knowledge Report, a series of detailed research reports on the state of the Arab world. The third report, *Arab Knowledge Report 2014: Youth and the Localization of Knowledge*, was gathered from workshops with more than 40 public administration experts (Mbrfoundation.ae, 2015c) and concluded that:

> Young people's possession of knowledge, skills, and values within an Arab vision to access the knowledge society, is the right way to realise the objectives of the integration of the youth in the process of transfer and localisation. This will help the Arab countries establish foundations of sustainable human development based on positive citizenship and social justice. (Mohammed Bin Rashid Foundation and United Nations Development Programme/Regional Bureau for Arab States, 2014, 211).

The report highlights a demographic challenge facing the Arab world – a "youth bulge" which under current economic and social conditions constitutes a possible threat to stability in the wider region, and suggests institutional reform as a way of translating this into "a human capital and real wealth that acts as a lever for moving towards the transfer and localisation of knowledge" (Mohammed Bin Rashid Foundation and United Nations Development Programme/Regional Bureau for Arab States, 2014, 210). In other words, the report makes a direct connection between the stability of the Middle East as a whole, and the establishment of knowledge-based centers such as those being set up in the GCC. Findings such as these will be of great interest to the rulers of GCC states and institutions and businesses across the region and beyond.

In October 2015, 28 Dubai-based organizations – including the MBRF – were brought together as the Mohammed Bin Rashid Global Initiative (MBRGI), gaining a collective Dh1bn (approximately $272,275,570) annual budget (Gulf News, 2015). It also brings together a number of the recurring elements of the GCC approach discussed in this paper – the eye-catching large numbers, statements of bold ambition, focus on a charismatic leader and interest in building a legacy through knowledge and development. Bringing together existing initiatives while improving funding, it also highlights the priorities of Dubai's ruler – literacy, health and innovation feature heavily in the MBRGI projects.

Focusing on the four areas of "combating poverty and disease, spreading knowledge, empowering communities, and entrepreneurship and innovation for the future" (Almaktouminitiatives.org, 2015) the MBRGI explicitly links the development of a knowledge-based society in the UAE with peace and prosperity in the region – and, being able to draw on the existing accomplishments of the individual initiatives, it can cite ready-made numbers on everything from books printed (3.2m) to entrepreneurs noted (23,000). The names of the initiatives are similarly eye-catching, including a "Museum of the Future" and "The UAE Drones for Good Award". How these initiatives fare in coming years will be a valuable indicator of the success of Dubai's, and the UAE's, approach to knowledge development.

Conclusion

National and regional needs are clearly always in mind as the GCC countries pursue, in their own ways, developments in education, science, technology, arts and heritage. The leaders of the GCC countries have more complex motivations than they are sometimes credited with – they are simultaneously trying to secure social stability in the face of regional unrest and their countries' position as regional (or even global) players in the context of global problems such as unpredictable oil prices and demographic change. The two problems are very much linked in their minds, and the solution to both appears to be the development of a knowledge-based society through large-scale investment in ambitious projects.

These projects are still in their infancy, and their success is still being gauged. Transnational knowledge brought into the GCC, funded by oil wealth, is a necessary part of the establishment of an educated local populace to advance the nations' ambitions in the long-term; the ultimate goal being that the next generation of GCC nationals will themselves be contributors to the worldwide knowledge exchange. This is sometimes characterized as a mission to make the GCC the center of a resurgent Islamic Golden Age of science and culture – validating decisions taken in the present by drawing attention to a past time in which Islam was a force for innovation. Not always taken seriously in the foreign press – where they are portrayed as initiatives taken primarily for the sponsor's glory, with no follow-through – the knowledge based centers are also up against some hefty odds at home, where many centuries have passed since the Golden Age and native interest in science lags behind the rest of the world. Everything discussed in this chapter happens in the context of, and in reaction to, this lag – many of the initiatives discussed here are very new, while the lack of interest in the sciences in particular is entrenched.

Writing in the science journal *Nature*, Zakri Report editors Nidhal Guessoum and Athar Osama noted that while OIC countries are home to nearly 25% of the world's people, they have contributed "only 1.6% of the world's patents, 6% of its academic publications, and 2.4% of the global research expenditure" (Guessoum and Osama, 2015). The new knowledge-based initiatives coming out of the GCC are intended as a remedy to this situation, and it is significant that many such initiatives look beyond the individual states' borders to the global Arab and Islamic population, offering a positive alternative to the status quo. As noted throughout this chapter, the Gulf states are investing substantial money in scientific research, in cultural initiatives and in social projects aimed at improving everything from literacy to entrepreneurship. They are also glamorizing science: the Emirates Mars Mission in particular functions on the idea that offering a huge – but achievable – goal within a relatively short timeframe is already kickstarting popular interest in science. To read the news in the UAE at the moment is to encounter on an almost daily basis a science-fictional idea made reality – from Mars probes and Masdar's self-driving cars to firefighters equipped with jetpacks to better fight fires in skyscrapers (Burgess, 2015). Of the latter, Jim Krane commented "This is [Dubai's] modus operandi, they like to take the first-mover advantage in just about anything" (Baraniuk, 2015). Indeed, the headline-grabbing approach is highly strategic from the point of view of a small, wealthy state seeking both investment and acknowledgement, and echoes the larger-scale projects already discussed. Guessoum (2014) comments in a Gulf News opinion piece that this initiative-seizing is valuable and quite deliberate, invoking both Sheikh Mohammed's comment that "epic challenges inspire and motivate us" and President Kennedy's famous rationale for going to the moon: "not because [it is] easy, but because [it is] hard".

The approach to developing knowledge-based institutions varies between GCC states. In accordance with different states' and rulers' ideals, particular areas may come to prominence at different times or in different forms. In the case of home-based versus branch universities, for example, in 2013 the UAE was home to 40 branch campuses, 27 of which were in Dubai alone, while Qatar had the next-highest number at 11, and Saudi Arabia only one (Swan, 2013).

Returning to the comparison of the role of patronage in Renaissance Italy, we return also to the key difference – that for all the competition to be the most innovative, dominant and influential, the GCC states are also interested in improving the lot of their citizens and stabilizing and enriching the larger Middle East. However, like the Italian princes, they do this out of enlightened self-interest – a stable region is one in which they can feel more secure. Each state, and each city, recognizes the need for an educated population and advanced institutions if they are going to be knowledge-based post-industrial and post-oil nations. Furthermore, the GCC is beginning to confront major critical challenges that will require worldwide cooperation, such as the reality of climate change – the prospect of global warming leading to summer temperatures in the GCC area that are "intolerable to humans" and possibly making summers in the Gulf unlivable by the year 2100 (Kermeliotis, 2015). This threat adds urgency to a desire to understand and prevent climate change, and to mitigate its effects. In November 2015 Dubai joined C40, a global alliance of cities working cooperatively to reduce emissions – touting its status as a "megacity" but still making a public commitment to an international collaborative effort and showing both a consciousness of the UAE's outsize carbon footprint and a desire to rectify it (Baldwin, 2015). It is perhaps these kinds of significant and cooperative endeavors that will be a legacy not only to the nations of the GCC, but to the world at large.

References

Adbookfair.com (2015a) KITAB and Abu Dhabi International Book Fair host the Embassies and Cultural Instututions. [online] Available at: http://tinyurl.com/hr3rchw [Accessed 10 January 2017].

Adbookfair.com (2015b) The spirit of Dickens haunts the Abu Dhabi International Book Fair. [online] Available at: http://tinyurl.com/zvuj8bz [Accessed 10 January 2017].

Adbookfair.german-pavilion.com (2015) German Pavilion – Abu Dhabi International Book Fair 2015 – Made in Germany. [online] Available at: http://tinyurl.com/jedqxq5 [Accessed 10 January 2017].

Aljazeera.com (2007) Dubai sets up $10bn education fund. [online] Available at: http://tinyurl.com/j6vhtq7 [Accessed 10 January 2017].

Almaktouminitiatives.org (2015) Mohamad Bin Rashid Al Maktoum Global Initiatives. [online] Available at: http://www.almaktouminitiatives.org/en/about/ [Accessed 10 January 2017].

Angre, K. (2015) PM Modi Rides Self-Driving Car in Masdar, World's First Zero-Carbon City. [video] Available at: http://tinyurl.com/zk3bsup [Accessed 10 January 2017].

Baldwin, D. (2015) Dubai joins climate change battle. [online] Gulfnews.com. Available at: http://tinyurl.com/jh5ooyd [Accessed 10 January 2017].

Baraniuk, C. (2015) Dubai wants to give its firefighters jetpacks. [online] Bbc.com. Available at: http://tinyurl.com/gn65txo [Accessed 10 January 2017].

Buid.ac.ae (2015) Programmes | The British University in Dubai. [online] Available at: http://www.buid.ac.ae/programmes [Accessed 10 January 2017].

Burgess, M. (2015) Dubai's firefighters get jetpacks to tackle highrise blazes (Wired UK). [online] Wired UK. Available at: http://tinyurl.com/h7zq77l [Accessed 10 January 2017].

Cannon-Brookes, P. (1984) The nature of museum collections. In: J.M.A. Thompson, ed., Manual of Curatorship, 1st ed. London: Butterworth.

Costa.co.uk (2015) Welcome to the Costa Book Awards – Costa Coffee. [online] Available at: http://www.costa.co.uk/costa-book-awards/welcome/ [Accessed 10 January 2017].

Diwc.ae (2015) About Dubai International Writers' Centre. [online] Available at: http://www.diwc.ae/en/about [Accessed 10 January 2017].

East, B. (2016) Making reading a sticky habit in UAE. [online] Vision.ae. Available at: http://vision.ae/articles/making_reading_a_sticky_habit_in_uae [Accessed 10 January 2017].

Emiratesliteraturefoundation.org (2015) Emirates Literature Foundation. [online] Available at: http://www.emiratesliteraturefoundation.org/index.html [Accessed 10 January 2017].

Emirateslitfest.com (2015a) Oxford University Press Short Story Writing Competition | Emirates Airline Festival of Literature. [online] Available at: http://tinyurl.com/hnwbsyn [Accessed 10 January 2017].

Emirateslitfest.com (2015b) Sponsors 2015 | Emirates Airline Festival of Literature. [online] Available at: https://www.emirateslitfest.com/sponsors [Accessed 10 January 2017].

Emirateslitfest.com (2015c) Qasidah par Coeur | Emirates Airline Festival of Literature. [online] Available at: http://www.emirateslitfest.com/qasidah-par-coeur [Accessed 10 January 2017].

Emiratesmarsmission.ae, (2015). Emirates Mars Mission – Space Vision. [online] Available at: http://www.emiratesmarsmission.ae/space-vision/ [Accessed 10 January 2017].

Emiratesmarsmission.ae (2015a) UAE makes plans to send unmanned probe to Mars by 2021. [online] Available at: http://www.emiratesmarsmission.ae/news/uae-mars-probe/ [Accessed 10 January 2017].

Emiratesmarsmission.ae (2015b) Emirates Mars Mission – Scientific Goals. [online] Available at: http://www.emiratesmarsmission.ae/scientific-goals/ [Accessed 10 January 2017].

Ghazal, R. (2015). Sharjah's planetarium offers a journey into our past and future. The National. [online] Available at: http://tinyurl.com/jvd6gwz [Accessed 10 January 2017].

Ghazal, R. (2015) Sharjah's planetarium offers a journey into our past and future | The National. [online] Thenational.ae. Available at: http://tinyurl.com/jvd6gwz [Accessed 10 January 2017].

Green, D. (2016) Sun-seekers to sergeants: UAE Year of Reading gets everyone on same page. [online] Vision.ae. Available at: http://tinyurl.com/jn7vp3r [Accessed 10 January 2017].

Guessoum, N. (2014) Main objective behind UAE's Mars mission. Gulf News. [online] Available at: http://tinyurl.com/hvdt8ne [Accessed 10 January 2017].

Guessoum, N. and Osama, A. (2015) Institutions: Revive universities of the Muslim world. Nature, 526 (7575), pp. 634-636.

Gulf News (2015) Dh1b Mohammad Bin Rashid Al Maktoum Global Initiatives launched. [online] Available at: http://tinyurl.com/hu3zhjv [Accessed 10 January 2017].

Hawking, S. (2011) Masdar City in Brave New World with Stephen Hawking ep03. [video] Available at: https://www.youtube.com/watch?v=jV5g5h5AyCc [Accessed 10 January 2017].

Kermeliotis, T. (2015) 'Deadly' heat waves predicted for Gulf states by 2100. [online] Aljazeera.com. Available at: http://tinyurl.com/npkj9lk [Accessed 10 January 2017].

Kingsley, P. (2013) Masdar: the shifting goalposts of Abu Dhabi's ambitious eco-city. Wired. [online] Available at: http://tinyurl.com/qh6hdt8 [Accessed 10 January 2017].

Leyne, J. (2007) BBC NEWS | Middle East | Dubai ruler in vast charity gift. [online] News.bbc.co.uk. Available at: http://news.bbc.co.uk/2/hi/middle_east/6672923.stm [Accessed 10 January 2017].

Louvreabudhabi.ae (2015) A Universal Museum - Louvre Abu Dhabi. [online] Available at: http://louvreabudhabi.ae/en/about/Pages/a-universal-museum.aspx [Accessed 10 January 2017].

Masdar.ae (2015) About Masdar Institute of Science and Technology. [online] Available at: http://tinyurl.com/zf4roc3 [Accessed 10 January 2017].

Masdar.ac.ae (2015a) Fast Facts. [online] Available at: http://www.masdar.ac.ae/about-us/fast-facts [Accessed 10 January 2017].

Masdar.ac.ae (2015b) Sponsored Research Centers. [online] Available at: http://tinyurl.com/jk3o6uy [Accessed 10 January 2017].

Mbrfoundation.ae (2015a) Welcome to Mohammed Bin Rashid Al Maktoum Foundation. [online] Available at: http://www.mbrfoundation.ae/en/Pages/default.aspx [Accessed 10 January 2017].

Mbrfoundation.ae (2015b) In line with directives of HH Sheikh Mohammed bin Rashid Al Maktoum MBRF launches 'Dubai Digital Library' initiative offering sophisticated electronic platform of Arabic books. [online] Available at: http://tinyurl.com/he54j39 [Accessed 10 January 2017].

Mbrfoundation.ae (2015c) Arab Knowledge Report. [online] Available at: http://www.mbrfoundation.ae/en/projects/Pages/AKR.aspx [Accessed 10 January 2017].

Mohammed Bin Rashid Foundation and United Nations Development Programme/Regional Bureau for Arab States (2014) Arab Knowledge Report 2014: Youth and the Localization of Knowledge. Arab Knowledge Report. [online] Dubai: Mohammed Bin Rashid Foundation and United Nations Development Programme/Regional Bureau for Arab States. Available at: http://tinyurl.com/j3dpzg9 [Accessed 10 January 2017].

Muslim World Science Initiative (2015) Report of Zakri Task Force on Science at Universities of the Muslim World. [online] London and Islamabad: Muslim World Science Initiative. Available at: http://muslim-science.com/science-at-universities-of-islamic-world-2/ [Accessed 10 January 2017].

Na.ae (2015a) Overview. [online] Available at: http://www.na.ae/en/aboutus/aboutna.aspx [Accessed 10 January 2017].

Na.ae (2015b) Corporate and Community Outreach Department. [online] Available at: http://tinyurl.com/hatg4gb [Accessed 10 January 2017].

Qatar Foundation (2012) HH Sheikha Moza on Qatar Science and Technology Park. [video] Available at: http://tinyurl.com/h9eogwn [Accessed 10 January 2017].

Qualey, M. (2014) Book Fairs Still Thriving in Riyadh, Muscat. [Blog] Arabic Literature (in English). Available at: http://tinyurl.com/ze4q6ob [Accessed 10 January 2017].

QSTP (2013) [YouTube Video] Qatar: IASPvisuals. Available at: http://tinyurl.com/zfu4vqb [Accessed 10 January 2017].

Qstp.org.qa (2015a) Qatar Science and Technology Park. [online] Available at: http://www.qstp.org.qa/ [Accessed 10 January 2017].

Qstp.org.qa, (2015b). Vision and Mission. [online] Available at: http://www.qstp.org.qa/home/about-us/vision-mission [Accessed 10 January 2017].

QSTP (2015) Qatar Science & Technology Park TECHtalks: a Glimpse into the Future of ICT and Health. [online] Available at: http://tinyurl.com/jf3ghp8 [Accessed 10 January 2017].

Rostamani, N. (2016) The UAE's Year of Reading will drive creativity - Khaleej Times. [online] Khaleejtimes.com. Available at: http://tinyurl.com/hox9s3m [Accessed 10 January 2017].

Scass.ae (2015) Planetarium Information. [online] Available at: http://www.scass.ae/en/about/Pages/about.aspx [Accessed 10 January 2017].

Sharjahart.org (2015) Sharjah Art Foundation. [online] Available at: http://www.sharjahart.org/about-us/about-us [Accessed 10 January 2017].

Sicc.ae (2015) Emirate of Sharjah. [online] Available at: http://sicc.ae/en/sharjah/emirate-of-sharjah/ [Accessed 10 January 2017].

Sorbonne.ae (2015) Sorbonne Abu Dhabi. [online] Available at: http://www.sorbonne.ae/about-us/sorbonne-abu-dhabi/ [Accessed 10 January 2017].

Swan, M. (2013) International branch campuses' key role in the UAE | The National. [online] Thenational.ae. Available at: http://tinyurl.com/p7bfvr4 [Accessed 10 January 2017].

TCA Abu Dhabi (2015) About National Library. [online] Available at: http://tcaabudhabi.ae/en/national-library/Pages/about-national-library.aspx [Accessed 10 January 2017].

WAM (2015) Hamdan bin Mohammed Heritage Centre becomes heritage strategic partner for Expo 2020. [online] Available at: https://www.wam.ae/en/news/emirates/1395274738559.html [Accessed 10 January 2017].

WAM.ae (2016) National Media Council announces first initiative for 2016 Year of Reading. [online] Available at: https://www.wam.ae/en/news/emirates/1395289764075.html [Accessed 10 January 2017].

WWF (2008) WWF, Abu Dhabi unveil plan for world's first carbon-neutral, waste-free, car-free city. [online] Available at: http://tinyurl.com/jfdhhe8 [Accessed 10 January 2017].

Zayednationalmuseum.ae (2015a) Zayed National Museum. [online] Available at: http://www.zayednationalmuseum.ae/ [Accessed 10 January 2017].

Zayednationalmuseum.ae (2015b) Architecture. [online] Available at: http://www.zayednationalmuseum.ae/architecture.html [Accessed 10 January 2017].

4

Interrogating Qatar's Urbanity as a Catalyst for Building Knowledge-Based Societies and Economies

Ali A. Alraouf

Introduction

Historically, Doha was a fishing and pearl diving town. Today, the capital is home to more than 90% of the country's 2.4 million people, the majority of whom are professionals from all around the globe (QSA 2015). Qatar's overall population has witnessed an unprecedented increase since 2010. With a current population of 2.4 million, of which only around 15 percent are native Qataris, the peninsula is just 11.571 square kilometers. Up to the mid-1960s, the majority of Doha's built environment was composed of traditional houses that represent local responses to the surrounding physical and socio-cultural conditions. During the 1970s, Doha was transformed into a modernized city. However, in the 1980s and early 1990s the development process was slow compared to the prior period due to the overall political atmosphere that resulted from the peaceful process of delegating power to Sh. Hamad Al Thani in 1995. Over the past decade, Qatar has become one of the world's major producers and exporters of Liquefied Natural Gas (LNG). The wealth produced by Qatar's oil and gas exports has generated a construction development boom in the capital, Doha, and the surrounding vicinity. This resulted in significant growth at all levels from urban development and infrastructure provision to cultural and educational facilities (Roberts, 2015). Three fundamental theoretical approaches with associated definitions will be investigated in this section. Defining the term "knowledge economy" and its impact on creating a new paradigm of urbanism is articulated. Then I will give a background on the topic of knowledge-based urban development in an international and regional context.

Post Globalized World: Cities Connecting the World

In the era of globalization, cities are regarded as arenas of accelerating change. Sassen (2001) refers to the four major interrelated functions that cities should perform in their attempts at becoming global. They should become "highly concentrated command points in the organization of the world economy; second, as key locations for finance and for specialized service firms, which have replaced manufacturing as the leading economic sectors; third, as sites of production including production of innovations, in these leading industries; and fourth, as markets for the products and innovations produced" (Sassen, 2001: 3-4). The third function implies the inevitable partial transformation of cities, within its globalization process, into knowledge-generating urban nodes. Castells (1996; 1998) argues that a new type of society is rising in our contemporary cities due to the consequences of the information revolution. Since the late 20th century, world cities have struggled to construct global identities. Castells argues that the global city is not a place but a process. The relational worlds of non-physical, functional, inter-locking networks of knowledge-intensive firms are to be conceptualized as 'spaces of flows', while the physical side of localized urban nodes of these same inter-locking networks are to be considered as 'spaces of places' (Castells, 1996). From a sociological point of view, Sassen (2001) argues that cities in the information age should be re-perceived as nodes of an immense network of cultural, commercial and political transactions. Given these points of view, it is obvious that the classical planning processes and design guidelines of contemporary cities need a substantial revision.

Sassen (2002) focuses on two key issues. First, she looked at how information flows have bound global cities together in networks, creating a global city web whose constituent cities become "global" through the networks in which they participate. Second, she investigated emerging global cities in the developing world - Sao Paulo, Shanghai, Hong Kong, Mexico City, Beirut, the Dubai-Iran corridor, and Buenos Aires. She showed how these globalizing zones are not only replicating many features of the top tier of global cities but are also generating new socio-economic patterns as well. These policies also focused on creating social amenities and communities to attract knowledge workers (Martin 2001; Chen and Choi 2004). The key factors in attracting knowledge workers to Knowledge and Creative Cities are social relationships and the quality of life in these cities (Mathur 1999; Leamer and Storper 2001; Robinson 2002; Santagata 2002).

The Concept of Knowledge and Creative Cities

To illuminate the connection between creativity, knowledge and city planning, Richard Florida's "Creative City" (2008, 2006, 2005, and 2002) theory is an essential tool for understanding how cities have become central to the achievement of competitive advantage. Therefore, the real economic competition today plays out among cities, not nation states as Florida rightly argues. An essential part of Florida's theory is the notion of a creative class, which is a body of talented people who would be attracted to a specific city. Florida's theory emphasizes three main factors: technology, talent, and tolerance. These three attributes are responsible for attracting the "Creative Class", contributing to the national economy, and establishing the requisite framework for sustaining competitive advantage. The creativity of cities is attributed to various factors, including openness to diversity, level of tolerance, and the peaceful coexistence of differences in a society. Levels of technological innovation and a population comprised of talented and innovative people are also believed to have a direct impact on the creativity, and hence, the development of a city. In his latest book, Florida (2008) asserts that people can now choose the cities in which they want to live, and this choice is shaped by the characteristics of specific cities.

A Paradigm Shift: From Resources to the Knowledge-Based Economy

The world has been moving through three distinctive economic stages. The first stage was resource dependent in that the prosperity of a particular economy was tied to the quantity and quality of resources available. The second stage resulted from the industrial revolution. The ability of a country to use technology to transform resources into new products which could then be exported to the rest of the world marked this type; finally, the knowledge economy has emerged as a dynamic process where knowledge and innovative ideas are used as vehicles for development. These shifts are essential for understanding the economic transformation underway in the Gulf (Alraouf, 2011).

Gulf Cities Transformed

The 21st century will be the era of cities, and cities will succeed or fail by their ability to understand that they, even more than countries or provinces, are the locus where the great economic, community and cultural issues will be played out. The countries in the Gulf region that form the Gulf Cooperating Council (GCC) share several common features, such as a high per capita GDP, monarchical rule, and abundant financial wealth generated from their dependence on oil and gas (Davidson 2008; Gray 2011). The GCC states have become more globalized and increasingly spend their rentier wealth more intelligently to develop their economies and societies, to diversify away from their strong reliance on oil, build new international images and roles for their cities and states, and even change the state's relationship with society (Gray, 2013: 2). Sixty years ago, today's glittering Gulf cities like Dubai, Abu Dhabi and Doha were tiny Bedouin or fishing villages with no real infrastructure or urban history, usually labeled "sleepy backwaters" of the Persian Gulf (Kamrava 2013). Kamrava (2013), states that there has been a steady shift in the regional balance of power in the Middle East away from the region's traditional heavyweights and in the direction of GCC states.

The last 20 years of urban development were characterized by a commitment to use oil revenues to allow primitive, small, and simple Gulf cities to rapidly upgrade urban infrastructures. During the first decade of the 21st century, major Gulf cities have emerged as rapidly growing knowledge economy localities. Unique for this region is the phenomenon of fast urban development. Observing the skylines of cities in the Gulf, or tracing their geographical boundaries during the last two decades, would prove what different researchers have characterized as an unprecedented pace of development (Davidson 2008; Gray 2011; Alraouf 2011; Gray 2013). More significantly, emerging cities in the Gulf have, in a very short timeframe, gained global importance in terms of connectivity with other prominent cities in global knowledge economy networks (Schein 2008).

Post Carbon Era: A Need for an Alternative Urbanism

The necessity of knowledge and creative cities in the Middle East has not been adequately explored. This chapter is an attempt to address this gap in the literature by assessing the principles that can be adopted to build knowledge-generative economies in a Middle Eastern context. For the establishment of contemporary creative cities in the Middle East, two integrated conditions are required. The process of constructing Middle Eastern knowledge and creative cities should be shifted from focusing on creating isolated and separated creativity centers to a process by which knowledge and creative networks are established and shared by different sectors of the community. Research is also a prerequisite of a knowledge and creative city. It requires an infrastructure that supports research, a component of development that has not been adequately prioritized in the Middle East (Alraouf, 2008). Hence, an initiative to prepare the grounds for a major transformation must be translated into strategy. This chapter will also argue that creative cities should be perceived as an opportunity for a sustainable model of growth and prosperity in the global knowledge-based economy. Therefore, emerging knowledge and cities in the Middle East should be seen as an integrative initiative that includes actors within both regional and global knowledge networks.

All Gulf States came to the conclusion that a focus on strategies for a post carbon paradigm is essential. The recent decline in oil prices from $100 to less than $40 suggests a swift change in the way Gulf States envision their future. Therefore, the objective of transforming into knowledge-based economies was declared as the backbone in all Gulf States' future visions including Bahrain 2030, Dubai 2030, Abu Dhabi 2032, Qatar 2030 and Riyadh 2025. Emerging knowledge cities in the Gulf, mainly Doha, Manama, Dubai, Abu Dhabi and more recently Riyadh, are moving from their previous status as oil producing economies to cities that celebrate education, research, and innovation, as they compete to attract knowledge workers (Alraouf, 2008). For this process to thrive, a different kind of urbanism is required. The needs and desires of migrant knowledge workers will have a profound effect on the built environment in these cities. Knowledge-based urban development (KBUD) can be an appropriate vehicle towards a new planning paradigm for Gulf cities. More specifically, establishing a creative or knowledge city implies its ability to affect spatial properties. New patterns of urban spaces should be articulated. The ultimate goal is to increase the innovation and creative capacity of cities based on a new set of knowledge patterns. The making of a creative city is a long and complicated process, but it is the path to follow for achieving sustainable urban development.

The Case of Doha, Qatar: An Interrogation of a Gulf Creative City

Historically, Doha, the capital of Qatar, was a fishing and pearl diving town. Currently, the city with its metropolitan borders is home to more than 90% of the country's 2.4 million inhabitants, the majority of whom are expatriates from other countries seeking a better working and living environment. In the past decade, Qatar has transformed itself into a major hub for numerous economic and cultural activities. Hosting a number of top-level sporting tournaments, culminating in the successful FIFA World Cup 2022 bid, has contributed to establishing Qatar and Doha as an emerging brand. Whether to counter regional economic competitors or to further tie Qatar to the economies of the world's leading countries, this brand is designed innovatively to address a range of security concerns; in short, Qatar is diversifying its dependencies (Roberts, 2015; Alraouf, 2016).

Furthermore, Qatar has become attractive as a place for foreign knowledge workers and, to use Florida's terminology, the creative class (Florida, 1996). Qatar's national vision for the year 2030 prioritizes the necessity of continuous social development in order to achieve a fair and safe society, while upholding human values and social welfare. The vision also emphasizes the imperative to strengthen its economy while continuing to satisfy the needs of citizens (QNV 2030). Maintaining the current growth rate of its urban population and improving quality of life is seen as critical to the future development of the country, as indicated in the national development strategy 2011–2016 (QSDP 2011). Qatar's economy has experienced three transformations. It has been 39 years since the British left Qatar and much has changed during the last four decades. Hydrocarbons have replaced pearls as Qatar's main source of revenue, bringing previously unimagined wealth to the former British protectorate (White, 2010). With a per capita income of $83,000, which is second in the world, the tribesmen of Qatar are building a new future for their country. In 2005, the Qatar Investment Authority (QIA) was established with the goal of reinvesting oil and gas revenues and building a diversified international asset portfolio. According to financial analysis done by RGE Monitor New York, Qatar has around $75 billion worth of investments outside the country (Roberts, 2015).

Strategically, Qatar has been on a multi-billion dollar spending spree to acquire assets across the globe. Yet, the country's leaders fully acknowledge the importance of development from within. So, while Qatar is spending a fortune on foreign shores, it is not forgetting to invest at home. As part of this development, the face of Doha will change immeasurably over the coming years. So far, Doha has succeeded in building a knowledge-generative infrastructure within the city. As clearly reflected in its future vision, Qatar's rulers acknowledge the shift towards a post oil paradigm (QNV 2030). More significantly, the leadership realized the reality of global competition between cities around the world. Hence, they acknowledge that in the context of international competition, new strategies for development have to emerge (Gray, 2013; Roberts, 2015). Undoubtedly, innovative architecture and dynamic urban planning models are excellent tools for establishing a global reputation. Doha, while investing in positioning itself on the map of emerging economies, was wise to use architecture and urbanism as a manifestation of a new era of planning and urban development (Alraouf, 2013; 2016). Governments in the Gulf also need to develop the "social infrastructure"-educational institutions and environments that foster innovation- on which the knowledge economy is based. People must be given a platform from which to innovate. Today there are an estimated 300,000 native Qataris, a number dwarfed by the ever-growing worldwide workforce residing in Qatar. In terms of such unique demographic structure, Qatar's experience can be seen as a positive example. Both of its knowledge economy mega projects, as will be explained, are creating interactive centers for knowledge production and transfer.

Knowledge-based urban development (KBUD) represents a new driving force in Gulf cities development; KBUD includes services, high tech parks, education, and is distinguished by the presence of knowledge workers or global citizens. In this section, an examination of KBUD projects in Doha will be conducted to evaluate current efforts to establish the city as a knowledge creative model within the Gulf and wider Middle East. Doha developed in an exceptionally short timeframe from a small fishing village to a global city attracting international knowledge economy firms. Until the late 1950s, it was a small town with no more than 20,000 inhabitants. The city developed on a reasonable pace throughout the following four decades until the end of 20th century. Finally, since the beginning of the new millennium, the city has become a global hub with more than 1.3 million inhabitants (United Nations. 2009). Doha has gained global significance through the growth of knowledge economy related projects. New spatial configurations along with the transformation of the city's built environment have contributed to the global attractiveness of Doha for knowledge economy investments, firms, and people. These changes have made Doha an attractive place for knowledge workers to live and to work (Roberts, 2015; Alraouf, 2016). Landry (2006) argues that even if we do not understand each other, we need to blend our differences. Hence, cultural diversity serves as an advantage for constructing knowledge and creative societies. While Doha's position is radically different from cities like Manama and Dubai when it comes to oil and gas reserves, Qatar leaders were convinced that the post-oil paradigm was a reality. Hence, moving from an industrial economy to a knowledge-based economy was an ineluctable transition on a global basis that required understanding and engagement. In line with the objectives outlined in Qatar Vision 2030, Sh. Hamad established the Qatar Foundation in 1995, a non-profit organization headed by Her Highness Sheikha Mozah Bint Nasser Al-Missned, to pave the way for the country's shift to a knowledge-based economy (White, 2010; Roberts, 2015). The Qatar Foundation is a comprehensive and dynamic knowledge structure concerning all educational strata, from primary to higher education. More importantly, QF sponsors creativity and innovation forums, a leadership academy, a sports academy, research centers, intellectual debates, and state of the art conference facilities. For the sake of this chapter, I will focus on some specific projects affiliated with the Qatar Foundation. These projects are Education City and Science and Technology Oasis. I will use the two projects with other projects from outside QF to evaluate Qatar's efforts to construct its identity as a knowledge and creative city within the Gulf's emerging urbanism.

The Vision, the Mission and the Adopted Strategies

This section covers the three main strategies adopted by the country's leaders to create a new sparkling pearl amid the Gulf States. Roberts (2015) argues that rarely has a state changed its character so completely and in so short a period of time. Previously content to play a role befitting its small size, Qatar was a traditional, risk-averse Gulf monarchy until the early 1990s. A bloodless coup in 1995 brought to power an emerging elite with a progressive vision for the future.

Extended Global Investment Arm

Oil and gas reserves have made Qatar one of the most rapidly growing economies in the world. Qatar became the world's top liquefied natural gas (LNG) exporter in 2007. Qatar's copious supplies of LNG allowed its small population to experience the globe's highest per capita income (Richer 2009). However, as mandated by Qatar 2030, Qatar changed the classical slogan of different Gulf States from "earn locally, spend globally" to "earn locally, invest globally." The traditional method of international investment was based on moving oil revenues to foreign banks and stocks. Qatar is forging a new path by focusing on substantial global investment. Qatar pays close attention to the diversification of its economy. Although the oil and gas related sectors still constitute the majority of state revenues, varying yearly from about 50 to 70 percent, diversification is high on the agenda (Richer, 2009; Kamrava, 2013; Roberts, 2015). As a good example of this sensitivity, the country's Sovereign Wealth Fund (SWF), the Qatar Investment Authority (QIA), valued at approximately US $115 billion at the end of 2012, "does not invest in the Qatari energy sector" (Gray, 2013: 105).

There are some reasons for the success story of QIA. First, QIA was fully activated during the global financial and economic crisis of 2008. Seizing opportunities in a climate of crisis proved to be an effective strategy in global investment. QIA has invested heavily in Western markets since its creation in 2005. For instance, QIA either outright owns or has stakes in many international firms, including UK supermarket chain Sainsbury's, the London Stock Exchange, Barclay's Bank, London's famous department store Harrod's, the Parisian department store Printemps, and a major French soccer club, Paris Saint-Germain. Qatar also has large stakes in Volkswagen and Siemens of Germany, Banco Santander of Brazil, and the Agricultural Bank of China. Very recently, QIA decided to invest more than 600 million US dollars in developing City Center DC, a shopping and residential complex in Washington DC. In light of such a wave of investment, Gray (2013: 107) claims that QIA "was created either primarily or partly as a deliberate strategy of the Emir toward balancing out the fluctuations in energy rents and diversifying the economy."

Political Credibility: A New Global Mediator

Qatar is, as Kamrava (2013) explains, a "tiny giant;" although severely lacking in most measures of state power, it is highly influential in diplomatic, cultural, and economic spheres. In the process of reinventing itself as a potential Arab super power, Qatar focuses on gaining global credibility. The Gulf State tries to build a niche as a conflict mediator within the Middle East and Africa. Using a combination of its financial power and its increasing diplomatic credibility, Doha is winning global respect as an able negotiator. Financed by gas exports, Qatar diversified its foreign relations to include Iran and Israel and carved a leading role in international mediation (Roberts, 2015). In recent years, it has endeavored to establish a unique profile as a diplomatic mediator and peace negotiator in conflicts ranging from Lebanon to Yemen and Darfur and, most recently, the Arab Spring countries. Qatar's prominent role in the Arab Spring follows a similar pattern, yet the gamble it is taking in supporting elected presidents and people's choices proved to be of a high cost. Qatar used the Arab Uprisings as an opportunity to portray itself as a regional advocate for freedom and democracy rather than a challenge to be contained (Ulrichsen, 2014).

Qatar has a palpable desire to be a trendsetter at the forefront in all domains, whether it is education, media or diplomacy. Therefore, the Emir is expected to continue to give priority to mediation efforts, boosting Qatar's standing in diplomatic circles. Qatar's effective use of its subtle power, Kamrava (2013) argues, challenges how we understand the role of small states in the global system. Qatar's diplomatic triumph is a result of the clear vision that Doha is the ideal environment for regional conflict resolutions and constructive negotiation. Kamrava (2013) points to the changing nature of power in the international arena in general and its utilization by Qatar in particular, labeling the new form of power that Qatar has carved out "subtle power."

Branding Qatar via Doha's Emerging Knowledge-Based Urbanism

Architecture, urbanism and planning are used, as will be explained in the coming sections, to prepare the country for the post-carbon era. Such preparation includes creating an environment to attract knowledge workers and creative people from around the world. On a planning level, the new blueprint for the urban development of Qatar, titled "Qatar National Development Framework (QNDF)," is the outcome of Qatar's urban planning department and its focus group, Qatar National Master Plan (QNDF 2016). The main pillars of the framework suggest a new form of urbanism for Qatar articulated around planning for sustainable growth, compact settlements, transit oriented urban development, walkability, mixed used urban centers, and a positive public realm. Such new planning principles aim at transforming Qatari cities and municipalities into people-friendly places and spaces to create an attractive environment for its knowledge economy workforce (QNDF 2016). For instance, enhancing walkability within the West Bay development will transform the area from a composition of isolated towers and skyscrapers into a network of activities, which would enhance social integration, walkability, and urban connectivity.

More significantly, according to QNDF and all municipal spatial development plans, Qatari cities are planning to accommodate a number of mixed-use urban centers. The centers will be diffused across the city. All centers will provide essential community services and facilities, including publically accessible green spaces. To enhance the sense of walkability, the centers are designed around either a metro station or a public bus node to emphasize the transit-oriented development strategy adopted in all of Qatar. Additionally, the National Spatial Strategy (2032) and Metropolitan Doha Plan (2032) are providing a future vision for the Qatari Capital. The vision is founded on a multi-centered approach to urban development. Metropolitan Doha contains three capital city centers that include Downtown Doha, Airport City, and West Bay. The three centers are characterized by mixed-use development, which can guarantee a vibrant, attractive, and sophisticated public realm. Also, environmentally-friendly transportation options such as water taxis, water buses, and bicycling are part of the plan. All of these policies and strategies substantiate QNDF's main objective to spatially redesign the city in ways that promote sustainable urban living for locals and expatriates (QNDF).

As mentioned earlier, QNV 2030 asserts the value of attracting knowledge workers to reside in Doha. To better cope with the conditions of establishing a knowledge economy, the vision suggests revisiting the country's demographic structure. While the current population is approaching 2.5 million, the vision estimates same number for 2030. The rationale, as explained in the vision, stems from the fact that Qatar's population will experience qualitative and not quantitative change. At least 800,000 construction and infrastructure workers will leave the country by 2019/2020 when most of the mega projects are finalized. This 35% of the total population will be replaced by knowledge workers and a creative class who will work the following fields: education, research, innovation, arts, banking, and in all other aspects of knowledge-based development (QNV, 2008).

Knowledge-Based Urban Development (KBUD)

Significant investment has been made in knowledge-based urban development in the country during the last decade. Architectural manifestations of the trend toward a knowledge-based urbanism around Doha include iconic projects like Education City, Qatar Science and Technology Park, National Library, Qatar National Museum (QNM), and Museum of Islamic Art (MIA). To learn more about these projects, I would start with the two museums, the MIA and the QNM, as they function to re-envision Doha as the new arts capital in the Middle East. MIA opened its door to visitors in December 2008. Inspired by the geometry of Ibn Tulun Mosque in the heart of Islamic Cairo, the museum was designed by the prominent architect, I.M. Pei. The museum is a contemporary representation of generative architecture produced from applying three dimensional geometrical complexities. MIA houses one of the world's most extensive collections of Islamic art and serves as the cornerstone of a monumental effort by Qatar to transform itself into the arts hub of the Middle East (Alraouf, 2016b). Qatar Museums (QM), which administers the MIA and other Qatari museums, has its own cultural development strategy, which is to turn Qatar into global and regional hub of culture and communication. Marking the next stage of its program to develop Qatar into a hub of culture and communications for the Gulf region and the world, the Qatar Museums (QM) revealed its plans for a new National Museum of Qatar, as expressed in a striking and evocative design by Pritzker Prize-winning architect, Jean Nouvel. The project is representative of general knowledge-based urban development trends in Doha and other Qatari cities. The first Qatar National Museum was opened on the June 23rd, 1975 and became the first official museum inaugurated in Qatar (Al-Mulla, 2014). Unlike the MIA, which was situated on a man-made island in the bay, the QNM is at the south end of Doha's Corniche. Whereas the old museum was located in the heart of the city, QNM will be the first monument travelers arriving from the airport will set their eyes upon. Jean Nouvel, in his design of the new QNM, provides an excellent example of the reimagining of existing museums. Unlike Pei, who was inspired by Muslim architecture for inspiration, Nouvel's design is responsive to the physical and environmental properties of the site and its proximity to the Gulf shoreline. The organic design allowed the new museum to embrace the old museum and, thereby, heighten its value, emphasizing the original museum's legacy. The construction of the new museum in and around the old one gives a message of integration and appreciation rather than separation and discontinuity between old and new. Precisely, the desert rose, a unique feature of the Gulf coast, made a huge impact on the architect's quest for a local natural element to be reinterpreted geometrically in the formal composition of the museum (Alraouf, 2016b).

Museums can be directly connected to the lifelong process of learning. The Museum's education programs are housed in a wing to the east of the MIA main building across a fountain courtyard. The Education Wing includes the Museum library, classrooms, workshops, study spaces, and technical and storage facilities. The MIA education center opened for the public in October 2009. It was established as a core component of the museum since the stress on its educational role was emphasized from the early stages of design. The center provides a variety of activities including public lectures and interactive workshops that are geared toward enhancing cultural life and raising the local community's awareness of art and cultural issues (Alraouf, 2016b). Becoming a regional center for education, knowledge, and culture is the aspirational identity for Qatar. The two museums, MIA and QNM, are designed and planned to engage the community through various activities, which transcend the typical role of the museum as a mere container of art works. Muslim administrators hope that these strategies for engaging the public will encourage more visitors to attend. The urban scenarios suggested by the design of the two museums will attract visitors via unique amenities like the open park at MIA with its regular concerts, bazaars, and recreational activities (Jodidio, 2009).

Another keystone in the vision of Qatar as a platform for KBUD is manifested in projects related to education, research, and knowledge dissemination. Education City (EC), located on the outskirts of Doha, has created a unique environment to support these efforts. EC hosts branch campuses of some of the world's leading universities, as well as numerous other educational and research institutions. Education City is envisioned as a hub for the generation of new knowledge: a place that provides researchers with world-class facilities, a pool of well-trained graduates, the chance to collaborate with like-minded people, and the opportunity to transfer ideas into real-world applications (Schein, 2008). While Education City is expanding dramatically, a major criticism for its pattern of development is based on the fear of transforming the city into a series of gated communities. To address these concerns, a deliberate effort is geared towards the internal and external integration of Education City with greater Doha. To achieve this goal some strategies were suggested including the adoption of a borderless campus, inviting the local community to use the city's facilities, and providing new amenities, which speak to the needs of the surrounding community and the residents of Qatar as a whole. I do serve as a member of the advisory board formed by Education City's "Capital Projects," the technical arm of the city to provide insights into the continuous process of integrating education city with the rest of Doha.

Similarly, Qatar's Science and Technology Park (QSTP) aims at turning Doha into a vibrant science and technology hub; attracting and retaining highly skilled employees are outlined in the Qatar Strategic Plan 2030. By the end of 2007, the first phase of Qatar Science and Technology Park was opened for business and populated with tenants. A massive, state-of-the-art convention center, another signature piece of architecture by Arata Isozaki, is already hosting prestigious global events. The Science and Technology Park provides facilities for commercial giants such as ExxonMobil and European Aeronautic Defense and Space Company. QSTP was established to provide a state of the art environment for local and global companies to develop and market technologically innovative products and services. QSTP's proximity to Education City facilitates research collaboration, innovation, and creativity (Schein, 2008; Kamrava, 2013).

Embracing Local Heritage and Global Urbanism

A well-balanced mixture of local and global approaches in planning and developing Doha's built environment highlights the emerging urbanity of the city. While the waterfront's development is characterized by iconic and signature buildings from the world's top architects, a number of projects are designed to preserve, conserve, and promote local architectural and cultural heritage (Alraouf, 2016a; 2016b). One interesting project, which brings people together to enjoy leisure time and also experience local culture is The Cultural Village (Katara). Katara is situated on Doha's eastern coast at West Bay, near the Qatar International Exhibition Center. Katara's concept is to create an environment suitable for nurturing cultural and innovative activities in the country by serving as a cultural hub and meeting place and to raise public cultural awareness through festivals, exhibitions, forums and other cultural events. Another unique example is Souq Waqif, originally a weekly market for local Bedouins. The souq acquired its name "Waqif," which means "standing" in English, because merchants stood up to peddle their goods. Coinciding with the emergence of modern Qatar, the souq developed to expand in space and in terms of the scope of activities that take place there. The souq's recent renovation is considered one of Doha's most successful projects. It has become one of Doha's most popular sites. After its renovation, Souq Waqif became a showcase for traditional architecture, handicrafts, and folk art. The souq authentically evokes Qatar's cultural heritage (Alraouf, 2016b). In contrast to modern construction styles that seem devoid of any cultural identity, its successful renovation highlights the nobility and wisdom behind the region's traditional architecture. This reality has made Souk Waqif imperative and the prime place to visit for locals, expatriates, and tourists alike. Souq Waqif was nominated for the prestigious Aga Khan Award for Architecture in the 2010. It has been described as a unique architectural revival of one of the most important heritage sites in Doha aimed to reverse the dilapidation of historic structures and remove inappropriate alterations and additions (AKAA Awards Report, 2010).

Concluding Remarks

In the Gulf, as well as elsewhere, one product of globalization is the increasing importance of knowledge and creative cities. As people and capital move more freely around the world, people will have more freedom in choosing where to live, leading to competition to attract global talent. This chapter has surveyed the strategies adopted by Qatar to compete in this global arena. Urban planning and architectural design constitute key elements of a strategy to craft a brand with global recognition. Qatari cities are being planned and designed to attract talented people, also known as knowledge workers, from around the world. Knowledge-based urban development should be perceived by Gulf cities as the most appropriate answer for their quest to move beyond economies dependent on the export of oil. A large part of the motivation behind these projects is the desire to establish Qatar as both a tourist destination and as a desirable place for knowledge workers to relocate.

References

Alraouf, A. 2016. "Dohaization": An Emerging Interface between Knowledge, Creativity, and Gulf Urbanity. Pp.: 47-68. In Katodrytis, George and Syed, Sharmeen. Gulf Cities as Interfaces. Cambridge: Gulf Research Centre Cambridge.

Alraouf, A. 2016b. Museums as a Catalyst for a New Urban and Cultural Identity in Qatar Interrogating the Case of Museum of Islamic Art. In Exell, Karen and Wakefield, Sarina (eds.) Museums in Arabia: Transnational Practices and Regional Processes (London: Ashgate).

Alraouf, A. 2014. From Pearling to Skyscrapers: The Predicament of Sustainable Architecture and Urbanism in Contemporary Gulf Cities. In Sustainable Development, An Appraisal of the Gulf Region (Berghahn Books: Oxford & New York, 2014).

Alraouf, A. 2012. An Emerging Urbanity in Knowledge Economy Paradigm: The Case of Doha, Qatar. International Conference on Realigning Power Geometries in the Arab World. Leipzig University, Germany. February 24-26, 2012.

Alraouf, A. 2011. Assembling Identity for an Emerging Urbanity. Doha, Qatar, as a Creative City. The City: 2nd International Conference. "The City: Culture, Society, and Technology." Vancouver, British Columbia. Canada. May 4-6, 2011.

Alraouf, A. 2010. Constructing Future Utopia: A New Paradigm in Knowledge-Based Urban Development, The Case of Doha, Qatar. 12th Conference of the International Association for the Study of Traditional Environments (IASTE, UC-Berkeley) 2010. American University of Beirut, Lebanon. December 15-18, 2010.

Alraouf, A. 2010. From Knowledge to Creativity Economy, Qatar Urbanity Transformed. Third Knowledge Cities Summit (KCS) 2010. Melbourne, Australia. November 16-19.

Alraouf, A. 2008. Emerging Middle Eastern Knowledge Cities: the Unfolding Story. In *Knowledge Based Urban Development: Planning and Application in the Information Era*. Edited by Tan Yigitcaular & V. Koray. USA: IGI Global Publishing.

Anholt, S. 2009. *Places: Identity, Image and Reputation*. Palgrave Macmillan.

Castells, M. 1996. *The Rise of the Network Society: The Information Age: Economy, Society, and Culture, Volume I.* Oxford, UK: Blackwell Publishers.

Castells, M. 1998. *End of Millennium, The Information Age: Economy, Society and Culture Vol. III.* Oxford, UK: Blackwell.

Carrillo, F. 2005 "Knowledge Cities: Approaches, Experiences, and Perspectives". Butterworth-Heinemann

Davidson, C. 2008. *Dubai: The Vulnerability of Success.* N.Y.: Colombia University Press.

Dinnie, K. 2011. *City Branding: Theory and Cases*. Palgrave Macmillan.

Evans, G. 2009. *Creative Cities, Creative Spaces and Urban Policy*. Urban Studies, 46 (5-6).

Florida, R. 2002. *The Rise of the Creative Class*. New York: Basic Books.

Florida, R. 2005. *Cities and the Creative Class*. New York: Routledge.

Florida, R. 2006. *The Flight of the Creative Class.* New York: HarperCollins Publishers.

Florida, R. 2008. *Who's your city?* New York: Basic Books.

Gehl, J. 2010. *Cities for People*. Island Press.

Govers, R. and Frank G. 2009. *Place Branding: Glocal, Virtual and Physical Identities, Constructed, Imagined and Experienced*. Palgrave Macmillan.

Gray, M. 2011. "A Theory of 'Late Rentierism' in the Arab States of the Gulf", *Occasional Paper No. 7*, Center for International and Regional Studies, Georgetown University School of Foreign Service in Qatar, 2011.

Gray, M. 2013. *Qatar: Politics and the Challenges of Development.* Boulder, CO: Lynne Rienner Publishers,

Harvey, D. 2009. *Social Justice and the City* (Geographies of Justice and Social Transformation). University of Georgia Press, Revised edition.

Jodidio, P. 2009. *Museum of Islamic Art, Doha, Qatar*, New York, NY: Prestel.

Kamrava, Mehran. 2013. *Qatar: Small State, Big Politics*. New York: Cornell University Press.

King, A. 2004. *Spaces of Global Cultures, Architecture, Urbanism, Identity*. London: Routledge.

Kotler, P. 1993. *Marketing Places*. Free Press.

Landry, C. and Phil, W. (eds.). 2007. *The Intercultural City: Planning for Diversity Advantage*. London: Earthscan Publications Ltd.

Landry, C. 2006. *The Art of City Making*. London: Earthscan Publications Ltd.

Richer, R. 2009. Conservation in Qatar: Impacts of Increasing Industrialization. Center for International and Regional Studies (CIRS), Georgetown University, Qatar.

Roberts, D. 2015. Qatar: Securing the Global Ambitions of a City-state. London: C Hurst & Co Publishers Ltd.

Sassen, S. 2001. The Global City: New York, London, Tokyo. 2nd edn. Princeton, NJ, USA.

Sassen, S. 2002. Global Networks, Linked Cities. Brunner-Routledge.

Soja, E. 2010. Seeking Spatial Justice (Globalization and Community). University of Minnesota Press.

Qatar National Development Framework (QNDF). 2016. Published by Ministry of Municipality and Environment. Doha, Qatar.

QSDP (Qatar General Secretariat of Development Planning) (2011) National Development Strategy 2011-2016. Gulf Publishing and Printing Company, Doha, Qatar.

Thierstein, A. and Schein, E. 2008. "Emerging Cities on the Arabian Peninsula: Urban Space in the Knowledge Economy Context", in ArchNet-IJAR: International Journal of Architectural Research, Volume. 2, Issue 2, Archnet- MIT Design Media Lab, Cambridge, USA, PP. 178-195.

Ulrichsen, K. 2011. *Insecure Gulf: The End of Certainty and the Transition to the Post-oil Era.* London: C Hurst & Co.

White, A. 2010. *How Doha's Dollars are changing the World.* In Arabian Business. Vol. 11 issue 28. Pp: 26-34.

Wood, P. and Landry, C. 2008. *The Intercultural City: Planning for Diversity* Advantage London: Earthscan.

5

The Role of Universities in the Innovation Systems of the Gulf: Gateways to the World of Knowledge?

Erik Baark

States that have joined the Gulf Cooperation Council (GCC) have traditionally relied on rich oil and gas exports for economic development. In the last decade, however, many governments in the region have shifted to promotion of knowledge-based development strategies (Engelke, 2015; Hussain, 2013). This shift has encompassed the promotion of innovative industries and the establishment of more comprehensive educational infrastructures, including the expansion of higher education. Universities have been seen as an important component of this new strategy, and most countries in the region have attracted overseas universities to set up branch campuses or developed local universities in close cooperation with transnational networks of higher education and research (Ewers and Maleki, 2010). The transition to knowledge-based economies implies that value in the economies is generated with ***created resources*** (creative educated human resources and their innovative products or services), in contrast to the current dependency on ***inherited resources*** (natural fossil resources).

This approach clearly involves a two-pronged ambition. On the one hand, the universities provide opportunities for students to be trained locally to raise the skill levels of key sectors in the economy rather than having to go abroad for further education. On the other hand, universities can attract talented academic scholars (including scholars from the Arab world who have obtained degrees overseas) to return to the region and conduct R&D that can contribute to raising technological levels locally. These two ambitions are in many ways complementary, but the emphasis appears to have been placed on the first – the educational mission – rather than the second in most Gulf countries. Developing vibrant innovation systems in the countries of the GCC implies a need to mobilize a similar transformation of universities from institutions emphasizing the transmission of basic professional skills to broad-based, curiosity-driven and innovative higher education.

This paper will examine the conditions for a transformation of the role of universities in the Gulf countries and investigate how the emerging innovation systems in these countries can accommodate such development.

Theories of Industry-University Partnerships: Understanding Institutional Embeddedness in Innovation Systems

Since universities first emerged in Europe during the Renaissance, the combination of developing new knowledge with the learning and dissemination of knowledge has remained a fundamental task for advanced institutions, even if the balance has shifted from one purpose to the other in different periods (Martin, 2012). Throughout their history over the centuries, universities have relied extensively on the expansion of scientific publication and academic communication to fulfill their function in society, often acting as "gateways" for access to international scientific results or technological achievements. Nevertheless, the function of universities as key elements in the development of advanced innovation systems has gained more recognition and emphasis in Europe and America, leading to the expansion of university-industry linkages and the emergence of entrepreneurial research-based universities, such as MIT, Stanford, or Cambridge (see Mowery and Sampat, 2005). More recently, the model of university-industry transfer has been taken up by universities in East Asia (see Sharif and Baark, 2008). Nevertheless, a survey of 22 universities in North America and in the Asia Pacific Rim found that technology transfer activities in Pacific Rim universities occur at a much lower level than in American universities but are growing at much faster rates (Wong et al., 2002).

The literature on university technology transfer primarily focuses on two modes of output. On the one hand, many studies focus on patents/technology licensing (Thursby and Kemp, 2002; Shane, 2002). Others, on the other hand, analyze knowledge transfer in the form of the creation of university spin-offs, i.e. new ventures that rely on the results of R&D conducted at universities (Steffensen et al., 2000). Research collaboration among teams of scientists and/or researchers from the corporate sector is a topic that has attracted increased attention in recent decades, and frequently this type of relationship forms the basis of university-industry linkages. A recent review of the scientific literature in the field found that trust between partners is essential for the success of collaboration, and that firm learning is one of the most significant outputs (Bozeman and Boardman, 2014).

In the United States, new incentives for university–industry technology transfer were introduced with the Bayh-Dole Act of 1980, which provided a legal framework for US universities to acquire ownership of intellectual property generated with federal funding. Engaging more actively in university-industry technology transfer, the effects of this framework have been examined with reference to leading universities such as Columbia University, Stanford University and the University of California (Mowery et al., 2004). Another study provided a concise analysis of both the US and international experience in a comparative perspective (Mowery and Sampat, 2005). These studies came to the conclusion that although there was no doubt that the Bayh-Dole Act had created a new context for university-industry linkages, the actual implementation of university initiatives was shaped by the previous trajectory of university-industry relations at each university. In this sense, the act appeared to have reinforced –rather than revolutionized – existing trends in such relationships.

Nevertheless, efforts to develop a better understanding of the role of universities and research institutions have increasingly been conceptualized within the Triple Helix model of innovation-driven economic growth, in which government agencies and private businesses work closely in complex, interactive relationships (Etzkowitz and Leydersdorff, 2000; Etzkowitz, 2008). The Triple Helix concept also provides a comprehensive framework for studying the new role of universities in terms of ideal interactions. This perspective also refers to the tensions that emerge as core responsibilities for training human capital for the future development of the knowledge-based society conflict with various profit motives and drivers of a Mode 2 approach to research (Nowotny et al., 2001).

Another important conceptual development has been the focus on the "Third Mission" of universities (Laredo, 2007). The argument here is that, in addition to the two basic missions of teaching and research, entrepreneurial universities engage in the third mission of contributing to society by diffusing knowledge to the local community, thereby generating research and consultancy income and assisting regional development. Many universities in the United States, Europe and Asia are strengthening their third mission activities and are being strongly persuaded by national or local governments to make this transition. Nevertheless, it is also crucial to recognize that the success of third mission activities depend on both the specialization of higher education organizations and the characteristics of the local context (Sánchez-Barrioluengo, 2014). Indeed, indicators of the innovative capacity of regions in the European Union show a strong correlation with those of highly skilled human resources, technological infrastructure, and creativity. A recent study argues that "[k]nowledge created in universities and other institutes of higher education is essential for the development of the required skills and expertise; it also allows companies to have access to state-of-the-art laboratories and capable graduates. There is a growing interaction between the educational system and the private sector in generating basic and applied research." (Sleuwaegen and Boiardi, 2014). International experience suggests that the effectiveness of intermediaries in the innovation system depends to a considerable degree on the organizational set-up and responsiveness of intermediary units vis-à-vis industry partners (Siegel et al., 2003). It is necessary to understand technology transfer from universities to the commercial sector within its broader context of innovation systems and their institutional features (Carlsson and Fridh, 2002).

Such a framework needs to be situated within an understanding of the innovation system in question. In particular, there are specific elements of the innovation system that have an impact upon potential linkages, including the scope and effectiveness of the intellectual property regime and the availability of financial support (which includes the banking system and availability of venture capital funding as well as conduciveness of the local stock markets for promoting start-ups). Such elements have often been emphasized in recent research on university-industry linkages in Asia (Sharif and Baark, 2011). The institutional characteristics are, to a large degree, determined by the structure of the innovation system and the development of new institutions.

Analysis of Innovation Systems

The concept of a national innovation system (NIS) has influenced the approach to science and innovation policy in most countries. However, in some developing countries the effectiveness of this approach has been greatly limited by a focus on the R&D system rather than the innovation system. While R&D may be a useful proxy for innovation in developed countries, it is not in developing economies, where R&D is not a major driver of development. In 'catching-up' economies, the key process is technological capability development, learning from the global stock of knowledge, and learning how to apply this knowledge. Capability development can be seen as a process of learning, and hence an innovation system is perhaps more usefully termed a learning system. However, it is important to recognize that all actors in the 'system' have to learn how to create value. In the case of firms, the market will select which firms create more value for customers and stakeholders, and those that are unable to change will fail. But what of public sector organizations? How do they learn to create value? What is the selection and improvement mechanism that drives continuous improvement? Moreover, what of policy makers and those who administer public programs – how do they learn?

Policies that explicitly promote and support capability upgrading in firms have been of great importance in many countries. However, such policies (and innovation policies in general) have little impact if a range of other policies is not consistent with these objectives. Thus, a reform of the innovation system that aims to encourage long-term investments in capability development requires close coordination of policies in areas such as trade, investment, education, and human resources that create an institutional framework for indigenous learning. The application of the NIS concept to innovation policy thus implies that the complete set of institutional factors influencing the creation and application of knowledge is surveyed in order to remove inefficiencies and obstacles to linkages in the economy and to dynamic learning processes for the various actors in the system.

A recent approach to developing an analysis of a national innovation system has become known as the functional assessment of NIS. There are actually several different analytical perspectives that underscore the importance of understanding the various functions performed in an innovation system. The first conceptual framework for a functional approach was launched by Liu and White (2001), who identified five essential functions of an innovation system, as illustrated by China's experience. This was subsequently elaborated by Charles Edquist into a perspective outlining ten different activities that were structured into four thematic categories: (i) the provision of knowledge inputs to the innovation process, (ii) demand side activities; (iii) the provision of constituents of systems of innovation; and (iv) support services for innovating firms (Edquist, 2005; Edquist and Hommen, 2008). However, the most well developed conceptual approaches were those proposed by a group of Swedish researchers led by Staffan Jacobsson that focused on the analysis of three structural components and seven dynamic functions of innovation systems (Bergek et al., 2008). The focus of such studies has been on specific technological sectors, such as the development of renewable energy, and not the entire national innovation system. Another elaborate approach has been conceptualized by a group of Dutch researchers who have analyzed innovation in the environmental sector and how it has been applied to environmental industry innovations in The Netherlands (Hekkert et al., 2007). In this paper, however, I develop a broader analysis of the functions of national innovation systems, using the specific role of universities – which are dominant new actors in the GCC innovation systems as a case study. I shall focus on seven functions:

- Resource mobilization: The extent to which an innovation system is able to mobilize essential resources for innovation such as funding, human resources, instrumentation, etc.
- Knowledge development: How key actors – such as universities, research institutes and firms – become involved in developing new knowledge.
- Market formation: To what extent the innovation system promotes new markets for innovative products, for example, through government procurement.
- Influence on direction of search: How innovation policy directs actors' research efforts towards areas of priority
- Entrepreneurial experimentation: How systems support new firms and ideas
- Legitimacy: Do cultural values and institutional frameworks in the innovation system encourage innovation?
- Development of external economies: How domestic and international networks support innovation.

The essential point is that analysis of an innovation system can focus on the way that the system supports a variety of functions – both in the supply of knowledge and innovative products and in the demand for innovation. In such analytical approaches, it is important to understand the extent to which actors will be able to perform their functions and jointly contribute to the innovation processes of the system. Moreover, a key issue is to what extent the institutional framework is providing adequate inducements or incentives for innovative activities – or whether existing institutions act as blocking mechanisms for such activities. Policies should, therefore, be designed to enhance incentives and remove blocking mechanisms – in other words, innovation policy assessment requires a wider scope of analysis, including a better understanding of the ways in which a range of policies (including economic policies, tax policies, trade and investment policies, IPR policies, educational policies, etc.) provides the right incentives for innovation.

For the purpose of this paper, we shall focus on whether or not – and in what ways – universities in the Gulf region are able to fulfill their functions in the innovation system. As mentioned in the section above, the fundamental function of universities is to ***educate*** a labor force to more advanced levels; that is, they are providing students with the basic knowledge required to undertake professional jobs. However, universities are also intended to help students in ***learning to learn***, meaning to develop dynamic and flexible learning patterns that allow them to further develop their skills and develop new knowledge. In this way, universities develop their second key function of research. They become more ***research oriented***, the role of doctoral studies in the portfolio of teaching expands, and research become important for the ***third mission*** of universities, namely to contribute to the application of new knowledge, entrepreneurship, and support for innovation in society. This transition has happened in many countries, with the effect that both advanced research universities and traditional colleges have increasingly come to underscore their functions as gatekeepers of innovation in the economy.

The Functions of Universities in Gulf Innovation Systems: Transition to Knowledge Production and Innovative Entrepreneurship?

There is a strong trend in the Arab Gulf States to emphasize the need for diversification in the economies as a way to meet the future challenges of over-reliance on oil or gas as export commodity and source of income. Such ambitions for promotion of new economic sectors and sources of income are clearly apparent in the various planning documents issued in recent years, and many policy initiatives in the region reflect these ambitions (Hvidt, 2013; Callen et al., 2014). One key initiative serving to diversify the economic structure in the Gulf countries has been a strong emphasis on the development of a knowledge economy – a policy that has figured prominently in the strategies adopted since 2005 in the GCC states (Hvidt, 2015; Orient Planet and Madar R&D, 2014). There are quite a few definitions for the concept of a knowledge economy, or a knowledge-based economy, but most of these emphasize the way the development and management of knowledge contributes to economic growth. A great deal of effort has gone into identifying appropriate indicators for knowledge economies, including efforts by international organizations such as The World Bank and OECD (Hussain, 2013). Most of the assessments attempt to capture both the extent and quality of the infrastructure for generating and communicating knowledge and the institutional context for learning and innovation that serves to enhance the performance of a knowledge economy (e.g., Hvidt, 2015; Ewers and Malecki, 2010; Engelke, 2015). In other words, while the majority of indicators for the role of knowledge and innovation in countries are based on quantitative indicators or ranking, the performance will often depend significantly on the quality of organizations, such as universities, on the role of interaction and cooperation in the innovation system and on the culture and incentives for innovation and knowledge-based entrepreneurship.

The available indicators for the progress of knowledge economies and innovation in the Arab Gulf States (see Table 1) show that their scores for economic incentive regime and ICT readiness are high, but the scores for education and innovation have been less spectacular, and for innovation, the scores are lower than world average. On the other hand, the information infrastructures have been expanded considerably during recent decades, with official figures showing a high level of ICT penetration in the population.

Table 1: Knowledge Economy Index Scores for GCC Countries, 2012

Country	Overall Knowledge Economy Index	Economic Incentive Regime	Innovation	Education	Information and communication Technology (ICT)
United Arab Emirates	6.94	6.50	6.60	5.80	8.88
Bahrain	6.90	6.69	4.61	6.78	9.54
Oman	6.14	6.96	5.88	5.23	6.49
Saudi Arabia	5.96	5.68	4.14	5.65	8.37
Qatar	5.84	6.87	6.42	3.41	6.65
Kuwait	5.33	5.86	5.22	3.70	6.53
World Average	5.12	5.45	7.72	3.72	3.58

Source: Hvidt (2015)

The diversity in these rankings is indicative of the limited linkage between higher education and innovation in the region that has been noted in a range of articles and book chapters in recent years, where one report notes that the hurdle of adult literacy rates is less serious in GCC countries (Orient Planet and Madar R&D, 2014). All the GCC countries have upgraded the educational infrastructure for several decades and have emphasized the expansion of opportunities for higher education in recent years. The growth of the number of universities in the Gulf States is reflected in Table 2 below.

Table 2: Universities in the Arab Gulf States, 2001-2008

Country	2001/02	2003/03	2003/04	2007/08
Bahrain	8	9	10	15
Kuwait	6	6	6	6
Oman	4	4	4	5
Qatar	3	4	5	7
Saudi Arabia	8	8	11	25
United Arab Emirates	8	8	8	15

Source: based on table in Abouammoh (2010)

The growth in terms of the number of universities in some countries – especially the United Arab Emirates and Saudi Arabia – appears to have been related to the expansion of the private sector (Davidson, 2010). The expansion of the university sector has been impressive, but the quality of education in the region is still criticized as falling behind international standards (see Barber et al., 2007; Kirk, 2015). For this reason, few universities in the region have made it to the higher ranks in the international surveys such as the Shanghai Academic Ranking of World Universities. King Saud University has the highest ranking of 197, while three other universities from Saudi Arabia are ranked in the 200-500 range. Nevertheless, the problem of linking universities to the innovation system persists, and the reasons for this include many aspects of the innovation systems and institutional framework for higher education (Akili, 2012).

It can be difficult to obtain hard data about details of innovation and scientific publications in the innovation systems in the Arab Gulf States. The only innovation survey published for Arab Gulf States that I have been able to access is a firm-level survey carried out in Abu Dhabi (Anderson, 2011). On the other hand, a source of quantitative data for outputs of innovation (scientific papers and patents) in the GCC states is Tseng (2014). It should be acknowledged from the outset that there is considerable variation in the conditions and experience of innovation in the Gulf States, but given the limitations of space, I will provide an overview of general trends together with a few selected details of cases from the region that illustrate these trends.

Resource mobilization: The aspect of resource mobilization is probably the area where the innovation systems of the Gulf States have witnessed the greatest progress in recent decades. In particular, there have been concerted efforts to raise financial resources for advanced education at several levels of society, not least by the establishment of universities. Financial resources have also been allocated to research infrastructure such as research institutes and science & technology parks or incubators. In fact, the development of major infrastructure in support of innovation has been a dominant approach among GCC policy makers, resulting in specialized cities that should foster high technology clusters for economic development, such as the Abu Dhabi Masdar City, the Dubai Healthcare City, and the Qatar Education City (Khodr and Reiche, 2012).

Universities have contributed – and should continue to contribute even more – to the mobilization of human resources. The majority of universities in the Gulf States have emphasized the professional or vocational training of skilled people who can fill important positions in government or industries, including the core industries associated with oil and gas exploration. A recent survey has revealed that more than 50% of the employees in both multinational corporations and locally owned firms in the United Arab Emirates and other GCC countries were university graduates. Many of these employees were expatriates, especially in UAE, but a university education was clearly important for the local employees too – especially in the sectors of resource-based manufacturing, trade and communications, and in construction and engineering (Ewers, 2013: 130-31). With the establishment of universities that emphasize research, post-graduate teaching, and supervision of PhD students in science and technology, such as King Abdullah University of Science and Technology (KAUST) in Saudi Arabia and Hamad bin Khalifa University (HBKU) in Qatar, the provision of highly skilled researchers will also become a critical function of the higher education sector.

Knowledge development: Many of the innovations that are currently implemented in the Arab Gulf States derive from external sources of knowledge, mostly through collaboration with people in the US (Tseng, 2014). A key source of new knowledge in advanced industrialized economies are public and private universities, but the universities in the GCC countries have only recently started to emphasize research (Royal Society, 2014) The result is that the number and level of scientific publications in many of the Gulf States have lagged behind international trends. Among the Gulf States, Saudi Arabia now has a substantial output of 7261 scientific publications, followed by the 1225 publications from the United Arab Emirates, while Kuwait, Qatar and Oman publish around 500 each. It is likely, however, that the size of knowledge output will increase dramatically with new investments in R&D in the region.

There are few public research institutes in the region, so this important source of development of new knowledge is rather weak in the Arab Gulf States. Therefore, a few large firms in the oil and gas sector will usually be the primary source of new knowledge for innovation. However, the economies that are less resource-dependent, such as the Emirates of Dubai and Bahrain, show that manufacturing and service firms have become more important for the creation of new knowledge (Anderson, 2011).

Market formation: The demand for innovation is primarily related to local markets for consumer goods and services and the requirements of international markets related to oil and gas exploitation or manufacturing related to these sectors (such as petrochemical industry). This has led to the emergence of a limited number of clusters in specialized industries, including those that rely on Free Trade Zone status (Figlioni, 2008). Nevertheless, the economies still lack diversification, which reduces the local market for products and services outside the oil and gas sectors; consequently, the economies are also characterized by limited export specialization (Callen et al., 2014). A well-recognized approach to the promotion of local markets for innovative products and services is to provide favorable incentives via major government-financed projects and public procurement. The GCC states have often engaged in major public procurement projects during the expansion of sectors such as oil, gas, and petrochemicals. To some extent, it still plays a crucial role in the construction of physical and ICT infrastructure, but so far, there is little evidence that government procurement has been utilized to support domestic innovation. In this sense, a vital element of market formation for innovative products and services utilized in advanced industrialized economies has not been apparent in the Arab Gulf States (Mahroum and Alsaleh, 2012a).

Entrepreneurial experimentation: Innovation is closely related to entrepreneurship, both in the sense that innovators need to be creative and enterprising in their search for new knowledge and in the sense that the commercialization of innovations is usually implemented through the establishment of new firms. In the Gulf States, the creation of new high tech firms has been supported for many years, but this particular function of innovation systems comes up against several institutional and cultural challenges (Schroeder, 2015). One institutional barrier is what has been called the neo-patrimonial political economies and the dominance of state-owned enterprises and family-owned firms in the economies of the Gulf (Osman, 2015). This has resulted in the subsequent weakness of the private sector and small and medium-sized enterprises (which only constitute 15-30% of all firms in GCC economies). The availability of attractive public sector jobs for most local males has also introduced a bias in the majority of Gulf societies that impedes private sector entrepreneurship. According to the consulting firm, Booz Allen Hamilton, the growth of a sustainable private sector in GCC states require major structural transformation in five key areas (Booz Allen Hamilton, no date). Interestingly, the effect has then been that innovative entrepreneurship among highly educated women has become a recognizable phenomenon (Engelke, 2015). It has also been pointed out in a recent report on the UAE that major steps forward have been taken to improve the enabling environment for transfer and localization of knowledge as a means to improve the conditions for youth entrepreneurship (UNDP, 2014).

Another crucial institutional issue is the labor market bias created by the role of expatriates. In most of the Gulf States, expatriate human resources are vitally important for the functioning of key economic sectors, especially in private industry and services. Expatriates tend to be segmented into highly trained professionals from the US, Europe or other Middle Eastern countries, and skilled labor that often arrive from Asia. This phenomenon has a tendency to reduce the participation of local nationals in private sector entrepreneurship (Osman, 2015; Ewers, 2013).

The function of providing entrepreneurial talent to the economies and inspiring the creation of new firms has not traditionally been important for universities, but again, this is likely to change in the future. Several of the Gulf States have set up special facilities and infrastructure to support entrepreneurial new high technology firms. Good examples are the Qatar Science and Technology Park, the Masdar Institute of Science and Technology in Abu Dhabi, Dubai's Techno Park, or the Bahrain Technology Park. The question remains, however, whether local universities are able to and motivated to contribute significantly to this function of entrepreneurial development (Akili, 2012; Jaruzelski et al., 2012).

Legitimacy: The institutional factors referred to in the discussion of entrepreneurship above reflect a range of social values that frame attitudes to innovation in the Gulf States. Paramount among these attitudes is the public support that innovation has received from leadership in all the Gulf States in recent decades – an approach that derives its rationale from a clear understanding of the need for reform – and which has been expressed in statements about national strategies and in the allocation of financial resources. However, in the population at large, there exist attitudes engendered by the rentier economy privileges that are problematic for the legitimacy of innovation and entrepreneurship activities (Hvidt, 2015: 16-18). Hvidt argues that the incentives to acquire and apply knowledge for entrepreneurial private sector jobs are, therefore, limited, in particular affecting the male population. This does not mean that young people in GCC countries do not have an interest in entrepreneurship at all, or that the governments have not expressed strong support for entrepreneurship, but it means that there are internal contradictions of the rentier state and structural factors which tend to discourage people from entering full-scale entrepreneurial ventures (Ennis, 2014).

Development of external economies: The functions of international networks are probably among the best in the Gulf States, which have for a long time relied on the transfer of knowledge from advanced industrialized countries. However, much of the knowledge has been acquired – rather than assimilated – in the key industries related to oil and gas exploration, and it has been argued that this approach has limited the development of local learning networks (Ewers, 2013). To some extent, the reliance on expatriate labor for much of the professional and technical work has further exacerbated this trend.

The function of universities as a gateway to external sources of knowledge is extremely important, and the Gulf States have already benefitted from the growth of branches set up by international universities (Wilkins, 2011). In addition, this function is expected to mature for all universities in the region in the future. There is no doubt that the development of higher education represents the apex of the educational reforms that the Arab Gulf States need to implement in order to promote the transition to knowledge societies (Davidson and Mackenzie-Smith, 2008; Wiseman et al., 2014).

The above discussion indicates that there exists a significant need for upgrading and strengthening of the various functions of the innovation systems in Arab Gulf States. Clearly, universities could play an important role in such efforts. The transition of universities to a central position in the emerging innovation systems would enable them to contribute substantially to the ambitious knowledge economies envisaged in strategic policy documents. Universities are an indispensable component of this process.

Conclusions and Policy Proposals

In this exploratory paper, I have examined the shifting role of universities in the emerging innovation systems of the Arab Gulf States and analyzed the motivations and achievements of the expansion of the Third Mission in terms of transfer of knowledge among universities. However, such a mission is of crucial importance for Gulf universities because of their dominant position in the development of knowledge economies in the region.

The innovation systems in the Gulf States can be investigated in terms of their essential functions, including resource mobilization, knowledge development, market formation and promotion of entrepreneurial capabilities. Although the function of resource mobilization has improved in many respects during recent decades, there is still room for improvement. For example, governments could do more to encourage existing industries to allocate substantial financial and human resources to R&D, and it will be crucial for the development of key sectors, such as renewable energy, that initiatives like the Masdar Institute of Science and Technology continue to gain major support.

One innovation system function that has not received adequate attention among policy makers in the Gulf countries is market formation. For instance, government procurement of renewable energy technologies could create market opportunities for this sector of the economy, which would complement the existing reliance of fossil fuel, and perhaps lead the way to stronger competitiveness in alternative energy supply in the future. There are elements of such a policy direction in the strategies adopted in UAE and Qatar, but I would argue that these elements need to be reinforced with even stronger incentives for firms in the market. Moreover, it can be argued that the remaining institutions characterizing rentier states undermine the efforts to legitimize innovation and entrepreneurship in Gulf societies. This hampers the crucial innovation system function of entrepreneurial experimentation and the expansion of the private sector in advanced sectors of industry and service in several Gulf States.

In conclusion, I would argue that universities might become essential contributing to the development of innovative capacity in the Gulf States and, thus, strengthen some of the innovation system functions that have been relatively weak in the Gulf economies. On the one hand, it would imply that the basic function of training professionals for government and industry positions should be enhanced by expanding the education of high level researchers and engineering personnel through the expansion of research postgraduate studies, for example. On the other hand, there is a key role of universities in strengthening the functions of interaction with international sources of knowledge and further developing this knowledge through advanced research to serve the processes of innovation in industry and services locally. This means that universities need to take more responsibility for assisting the application of innovative products and processes in society and instilling a culture that celebrates high technology visionaries and provides adequate incentives for the people whose ambitions can be pursued through private sector jobs.

References

Abouammoh, A. M. 2010, "The Role of Education: Reform Trends and EU-GCC Cooperation" in: C. Koch and L. Stenberg (eds.) *The EU and the GCC: Challenges and Prospects under the Swedish EU Presidency* Dubai: Gulf Research Center, p. 79-96.

Akili, W. 2012, "Academia Industry Collaboration in the Arab Gulf States: Divergence or Convergence?" Paper presented at the 2012 ASEE Annual Conference: http://www.asee.org/public/conferences/8/papers/3248/view (accessed 10 January 2017).

Anderson, T. 2011, "Natural Resource Dependency and innovation in the GCC Countries" in F. S. Nobre, D. Walker and R. Harris (eds.), *Technological, managerial and organizational core competencies: dynamic innovation and sustainable development* Hershey, Pennsylvania: IGI Global.

Barber, M. M. Mourshed and F. Whelan 2007, "Improving Education in the Gulf" *The McKinsey Quarterly*: http://tinyurl.com/jbfk59o (accessed 10 January 2017).

Bergek, A., S. Jacobsson, B. Carlsson, S. Lindmark, and A. Rickne 2008, "Analyzing the functional dynamics of technological innovation systems: A scheme of analysis" *Research Policy*, Volume 37, Issue 3, (April 2008), pp. 407-429.

Booz Allen Hamilton (No date) *Building a Diversified Economy: How GCC Nations Can Stimulate Private-Sector Innovation and Sustain Long-Term Growth*: http://tinyurl.com/h7kw8b8 (accessed 10 January 2017).

Bozeman, B. and Boardman, C. 2014. *Research Collaboration and Team Science: A State-of-the-Art Review and Agenda*, Springer.

Callen, T., R. Cherif, F. Hasanov, A. Hegazy, and P. Khandelwal 2014, *Economic /Diversification in the GCC: Past, Present, and Future* Washington: IMF Staff\Discussion Note SDN/14/12.

Carlsson, B. and Fridh, A.C. 2002, "Technology transfer in United States universities – A survey and statistical analysis", *Journal of Evolutionary Economics*, Vol. 12, No. 1-2, p. 199-232.

Davidson, C. M. 2010, "The Higher Education Sector in the Gulf: History, Pathologies, and Progress" in: C. Koch and L. Stenberg (eds.) *The EU and the GCC: Challenges and Prospects under the Swedish EU Presidency* Dubai: Gulf Research Center, p. 59-75.

Davidson, C. M. and P. Mackenzie-Smith 2008, *Higher Education in the Gulf: Building Economies, Politics and Cultures*. London: Saqi Books.

Edquist, C. 2005, "Systems of Innovation: Perspectives and Challenges," in Fagerberg, J., Mowery, D., and Nelson, R. (eds.) *Oxford Handbook of Innovation* Oxford: Oxford University Press, pp. 181-208.

Edquist, C. and L. Hommen (eds.). 2008. *Small Country Innovation Systems: Globalization, Change and Policy in Asia and Europe* Cheltenham U.K.: Edward Elgar.

Engelke, P. 2015, *Brainstorming the Gulf: Innovation and the Knowledge Economy in the GCC* Washington: Atlantic Council Brent Scowcroft Center on International Security.

Ennis, C. A. 2014, "Between Trend and Necessity: Top-Down Entrepreneurship Promotion in Oman and Qatar" *The Muslim World* Vol. 105 (January 2015), p. 116-138.

Etzkowitz, H. 2008, *The Triple Helix: University-Industry-Government Innovation In Action* London: Routledge.

Etzkowitz, H. and L. Leydesdorff 2000, "The dynamics of innovation: from National Systems and "Mode 2" to a Triple Helix of university–industry–government relations" *Research Policy*, Volume 29, Issue 2, p. 109-123.

Ewers, M.C. 2013, "From knowledge transfer to learning: The acquisition and assimilation of human capital in the United Arab Emirates and the other Gulf States" *Geoforum* Vol. 46, p. 124-137.

Ewers, M. C. and E. J. Malecki 2010, "Leapfrogging into the Knowledge Economy: Assessing the Economic Development Strategies of the Arab Gulf States" *Tijdschrift voor Economische en Sociale Geografie* Vol. 101, No. 5, p. 494-508.

Figliolini, A. J. 2008, "Free Trade Zones: The Development of Economic Clusters in the GCC" *Dubai: Center for Advancement of Human Resource Studies*: http://tinyurl.com/z5w4ph2 (accessed 10 January 2017).

Hekkert, M.P., R.A.A. Suurs, S.O. Negro, S. Kuhlmann, and R.E.H.M. Smits 2007, "Functions of innovation systems: A new approach for analysing technological change" *Technological Forecasting and Social Change*, Volume 74, Issue 4, (May 2007), pp. 413-432.

Hussain, A. 2013, "Evolution of Mutual Knowledge-Based Economy in Regional Integration: An Experience from the Cooperation Council of Arab States of the Gulf" *Journal of the Knowledge Economy* (February 2013) DOI 10.1007/s13132-013-0148-5.

Hvidt, M. 2013, *Economic Diversification in GCC Countries: Past Record and Future Trends* (Research Paper, Kuwait Programme on Development, Governance and Globalisation in the Gulf States) London School of Economics.

Hvidt. M. 2015, *Transformation of the Arab Gulf Economies into Knowledge Economies: Motivational Issues Related to the Tertiary Educational Sector* Doha: Arab Center for Research and Policy Studies.

Jaruzelski, B., C.N. Moujaes, R. Eltayeb and J. Hajd 2012, *Coherent Linkages: How to Foster Innovation-Based Economies in the Gulf Cooperation Council* Booz & Company: http://tinyurl.com/zw8hzgq_(accessed 10 January 2017).

Kirk, D. 2015, "Innovate or Replicate? Education Reform Initiatives in the Gulf Cooperation Council States" *The Muslim World* Vo. 105 (January 2015), p. 78-92.

Khodr, H. and D. Reiche 2012, "The Specialized Cities of the Gulf Cooperation Council: A Case Study of a Distinct Type of Policy Innovation and Diffusion" *Digest of Middle East Studies* Vol. 21, No. 1, p. 149-177.

Laredo, P. 2007, "Revisiting the Third Mission of Universities: Toward a Renewed Categorization of University Activities?" *Higher Education Policy*, Vol. 20, p. 441-456.

Liu, X. and S. White 2001, "Comparing innovation systems: a framework and application to China's transitional context" *Research Policy* Vol. 30, No. 6, pp. 1091-1114.

Mahroum, S. and Y. Al-Saleh 2012a *Demand-led Related Diversification: Green Construction in the GCC as a Prospective Case* Dubai: INSEAD Faculty & Research Working Paper 2012/34/IIPI.

Martin, B. R. 2012, "Are universities and university research under threat? Towards an evolutionary model of university speciation" *Cambridge Journal of Economics*, Vol. 36, p. 543–565.

Mowery, D.C., Nelson R.R., Sampat, B.N., and Ziedonis, A.A. 2004, *Ivory Tower and Industrial Innovation* Stanford: Stanford University Press.

Mowery, D.C and Sampat, B.N. 2005, "Universities in National Innovation Systems" in J. Fagerberg et al (ed) *The Oxford handbook of Innovation* Oxford: Oxford University Press.

Nowotny, H, P. Scott, and M, Gibbons 2001, *Re-Thinking Science: Knowledge and the Public in an Age of Uncertainty*. London: Polity Press.

Orient Planet and Madar R&D 2014, *Arab Knowledge Economy Report* Dubai: Orient Planet and Madar Research & Development.

Osman, D. 2015, "The State and Innovation – An Analytical Framework" *The Muslim World* Vol. 105 (January 2015), p. 2-23.

Royal Society 2014, *The Atlas of Islamic World Science and Innovation: Final Report* London: The Royal Society: http://tinyurl.com/js48ycv (accessed 10 January 2017).

Sánchez-Barrioluengo, M. 2014, "Articulating the 'three-missions' in Spanish universities", *Research Policy* Vol. 43, p. 1760-1773.

Schroeder, C.M. 2013, *Startup Rising: The Entrepreneurial Revolution Remaking the Middle East* New York: Palgrave Macmillan.

Shane, S. 2002, "Selling university technology: Patterns from MIT", *Management Science*, Vol. 48, No. 1: p. 122-137.

Sharif, N. and E. Baark 2008, "Mobilizing Technology Transfer from University to Industry: The Experience of Hong Kong Universities" *Journal of Technology Management in China*, Vol. 3, Issue 1, p. 47-65.

Sharif, N. and E. Baark 2011, "The Hong Kong University of Science and Technology: A Case Study in Entrepreneurial-University-Led Knowledge Based Economic Development" in P. K. Wong (ed.) *Academic Entrepreneurship in Asia: The Role and Impact of Universities in National Innovation Systems*, Cheltenham, UK: Edward Elgar Publishing, p. 135-164.

Siegel D.L., D. Waldman, A. Link 2003, "Assessing the impact of organizational practices on the relative productivity of university technology transfer offices: an exploratory study" *Research Policy* Vol. 32, p. 27-48.

Sleuwaegen L. and P. Boiardi 2014, "Creativity and regional innovation: Evidence from EU regions" *Research Policy* Vol. 43, p. 1508-1522.

Steffensen, M., E.M. Rogers, and K. Speakman 2000, "Spin-offs from research centers at a research university" *Journal of Business Venturing*, Vol. 15, No. 1.: 93-111.

Thursby J.G. and Kemp, S. 2002, "Growth and productive efficiency of university intellectual property licensing" *Research Policy*, 31(1): 109-124.

Tseng, C. Y. 2014, "Technological innovation capability, knowledge sourcing and collaborative innovation in Gulf Cooperation Council countries" *Innovation: Management, policy & practice*, Vol 16, No. 2: p. 212-223.

UNDP. 2014. *Arab Knowledge Report 2014: Youth and Localisation of Knowledge United Arab Emirates*: http://tinyurl.com/jpts8ce (accessed 10 January 2017).

Wilkins, S. 2011, "Who benefits from foreign universities in the Arab Gulf States?" *Australian Universities' Review*, Vol. 53, no. 1, p. 73-83.

Wiseman, A. W., N. H. Alromi, S. Alshumrani (eds.) 2014, *Education for a knowledge society in Arabian Gulf countries* London: Emerald.

Wong P.K., K. Allen et al. 2002, Survey of Technology Transfer and Wealth Creation (T2WC) Among APRU Member Universities*: Survey Report.*

6

Japanese Efforts to Build Smart Power in the Gulf

Isam R. Hamza

Introduction

It is no exaggeration to say that Japanese policy towards Middle Eastern countries has been reactionary for a long period of time. It is well known that relations between Japan and the Middle East have rested primarily on the import and export of crude oil and natural gas (LNG), which is a source of life, not only for industry, but also for many other energy-dependent activities in Japan.

Following the first "Oil Shock" in 1974, those relations showed a marked change. Japan was the most dependent country on Arab oil in the 1970s, where 71% of its imported oil came from the Middle East. To encourage the Japanese government to change its "non-involvement policy", the Saudi and Kuwaiti governments declared Japan a "non-friendly country" at the onset of the Oil Crisis in November 1973 – after the October 6th War or, as it is called in Japan, "the Fourth Middle East War". As a result, Japan received a 5% production cut in December of that year, causing panic in different sectors of the Japanese economy.

It was the first time in Japan's contemporary history that Japanese citizens waited in long queues at supermarkets, department stores and closed gas stations, to obtain sundry household items or a few liters of petrol. This event threw the daily life of the Japanese people into chaos, and threatened to drastically curtail the country's expected growth rate (ActionForex, 2004).

On November 22nd, Japan issued a statement advocating for Palestinian self-determination, asserting that Israel should withdraw from all of the 1967 territories. Additionally, Japan threatened to reconsider its policy towards Israel if the latter refused to accept these conditions. By December 25th 1967, Japan was considered an Arab-friendly state (Sirkhabi, 2015:92-96).

In response to the oil crisis, Japan embarked on a new policy towards the Arab countries, one that not only involved oil-based relations, but also a knowledge-based exchange. The implementation of Japan's knowledge-based policy in the Middle East has been intermittent over the last 40 years, with its success being dependent upon economic conditions and energy supply problems during that period.

Moreover, wars in the Gulf Area, and unstable conditions in the Middle East and Arab World, have always affected the Japanese approach to the Gulf. Japanese initiatives to develop human capital and foster knowledge-generative networks began with Egypt and were later extended to Saudi Arabia, the UAE, and Qatar. With the potential to expand to a variety of cultural, diplomatic, and economic sectors these initiatives are principally designed to strengthen and ensure the longevity of Japan-Gulf relations.

The Beginning: Egypt

Fortunately, one year before the Oil Shock, a special legal entity, supervised by Japan's Ministry of Foreign Affairs, was established in October 1972 under the name of "The Japan Foundation". Its objective was to promote international cultural exchange through a comprehensive range of programs (The Japan Foundation, 2015). Because of these Japanese initiatives, an agreement was reached with the Egyptian government, and the Department of Japanese Languages and Literature (hereafter The Japanese Department), was founded at Cairo University in Egypt in 1974. This marked the first academic unit specializing in Japanese Studies in an Arab country. Subsequently, The Japan Foundation has dispatched many Japanese professors and language instructors to The Japanese Department over an extended period of time.

In October 1974, approximately 200 students applied to the new Japanese Department. It was the first academic year following the war, and most of the student applicants (including the author of this paper) were eager to learn about the East, so a chance to gain further knowledge of Japan motivated them to join the new Japanese Department. Furthermore, in the early 1970s, the spread of Japanese electrical appliances, cars and other fine products in the Egyptian market promoted student curiosity in learning more about this rapidly developing country and its valuable experience in meeting the challenges of modernity. Strikingly, though, there was no motivation to join the Japanese Department as a means of securing employment with Japanese companies or national agencies dealing with Japan, simply because no such entities existed at that time.

The one and only visiting Japanese faculty member at the time selected 43 out of 200 students for the inaugural class (including the author of this paper), based on student examination results. Teaching conditions proved difficult, due to a lack of sources – textbooks, Japanese- English dictionaries, etc. Only one mimeograph was used that first year. In the following year, The Japan Foundation supplied the department with textbooks and dictionaries, and dispatched teaching staff specialized in Japanese linguistics, history and Arabic Studies. The department also reduced the number of entering students to 20, later increasing it to 25. Today, The Japanese Department has 12 faculty members with doctoral degrees (including the author), enabling them to participate in cultural and educational activities both inside and outside of campus, therefore ensuring a more robust vehicle for the exercise of Japanese smart power.

The author himself had an opportunity to participate as an advisor to the Japan International Cooperation Agency (JICA) Cairo Office for Educational Affairs, and was appointed Special Advisor for the Minister of Education, so that he could introduce the benefits of the Japanese model for the Egyptian educational system. He also was the first Egyptian to be nominated as head of The Japanese Department at Cairo University in 1998-2001, and was nominated again in 2007-2009. He served as Vice Dean for the Faculty of Arts at Cairo University until 2011, and during these terms, he struck academic agreements with more than 15 famous Japanese universities. He appeared on many TV programs, and wrote articles on Japanese culture, thought, politics and Arab-Japanese relations, trying to fulfill his expected role as an interlocutor helping explain Japan to the Egyptian public.

The Japan Foundation's efforts to disseminate Japanese learning and knowledge of Japan continued. In 2000, after a quarter of a century, it contributed to the establishment of a new Japanese Language Department at the Faculty of Foreign Languages (Al-Alsun) at Ain Shams National University in Cairo. And in September 2013, in spite of the exodus of native Japanese instructors from Egypt (due to the uneasy situation that Egypt faced), a new Japanese Language Department was established in Aswan National University in the south of Egypt. It began accepting students in September 2014, and utilized visiting Egyptian faculty from Cairo University.

There is also a Department of Japanese Language and Translation at Misr University for Science and Technology (MUST), a private university since 2005. Established without the support of The Japan Foundation, the department was founded by the individual efforts of an Egyptian professor (the author) from Cairo University, and is now run by his colleagues.

In October 2003, the Japan Foundation was recognized as an independent administrative institution, based on a Japanese Government endowment of 78 billion yen. Its activities were financed by annual government subsidies, investment revenues and donations from the Japanese private sectors.

Although Egypt is not a member of OAPEC, it is considered one of the most important countries in the region for Japan. It ranks third in GDP, after Saudi Arabia and the UAE (in 2012), where 20% of Egypt's GDP relies on oil and natural gas production (Hosaka, 2015). It is also worth mentioning that Egypt's population is the biggest in the Arab world. Moreover, it possesses the Suez Canal, where most Japanese car exports destined for Europe must transit. For these reasons, instability in Egypt would definitely give the Japanese economy a severe shock.

Egypt also is the cultural and educational center of the Arab World. It plays a great role in the fields of media and entertainment in the region. Egyptian universities, like Cairo University, Ain Shams University, AUC (American University in Cairo) and Al- Azhar

University, receive thousands of students from many Arab and African countries every year (Hosaka, 2015).

The Department of Japanese Language and Literature in Cairo University has not only accepted Egyptian students, but also other students from neighboring Arab countries, such as Syria, Sudan, Jordan, and the UAE. Some of the Egyptian and Arab graduates have had the opportunity to complete their postgraduate studies in national Japanese universities through the Ministry of Education, Culture, Sports, Science and Technology (MEXT) scholarships. After obtaining their doctoral degrees in Japan, the first such group returned in the early 1990s. Many members of this first cohort were hired to serve as faculty in the department. During that time, the Japanese Government also sent Japanese professors, lecturers and language instructors to Cairo University through The Japan Foundation programs.

Transnational Experts in KSA

By the beginning of the 1990s, the situation in the Gulf area had become unstable, as the "Gulf War" (2 August 1990-28 February 1991) had just broken out on the heels of the near decade-long war between Iran and Iraq (1980-1988). The Gulf War damaged Japan's economy, and threw Japanese diplomacy into disarray. The rise in oil prices caused an unexpected spike in inflation at a time when the Japanese economy was facing the so-called "Bubble Era". Japanese diplomacy also faced external pressures to participate in the multinational military forces being dispatched to the region. However, the ninth article of the Japanese constitution effectively prevented the government from making any military contribution (Nakanishi, 2011).

At that time, Foreign Minister Nakayama paid a short visit to Saudi Arabia and Oman. This visit played a direct role in helping Japan review its diplomatic identity, which led Japan to rededicate itself to promoting peaceful economic relations with the region. At that time, "globalization" was the slogan raised by the government of Japan in trying to promote a new world order. As part of that slogan, The Japan Foundation contributed to the establishment of Japanese language programs in concert with the Department of Asian Language and Translation at the Faculty of Language and Translation, King Saud University in Riyadh in 1994. This program is the only one in the GCC that offers a bachelor's degree in Japanese Studies and, as such, is expected to shoulder the responsibility of teaching the Japanese language in the region. One Japanese faculty member was dispatched to launch the program, and, for the first time, two Egyptians were nominated as full-time faculty members. These two Egyptians graduates of the Department of Japanese Language and Literature, both of whom obtained their doctoral degrees at Japanese universities, demonstrated the success of Japan's early efforts to develop the human capital necessary to realize its regional smart power strategy.

Since 1994, there have always been two Egyptians and one Japanese instructor running the program. Initially, the study of Japanese in Saudi Arabia was limited to male students but has, in recent years, enlarged to include female students too. Japanese language courses were offered in 2010-2011at Dar Al-Hekma Women University in Jidda. In 2011, Japan's Ministry of Foreign Affairs dispatched female language instructors to some Saudi women's universities and junior colleges, such as Princess Nura, Dar Al-Hekma, Effat, and the female section at Prince Sultan University, which established short-term courses for the study of the Japanese language. Also in 2011, Saudi female instructors started to step onto the platform of Japanese teaching at the female sections of King Abdel Aziz University and Um- Alqura University (Embassy of Japan in Saudi Arabia, 2015).

As a result, "The Fifth Japanese Language Speech Contest" held in Riyadh in 2012 had a slot for female contestants. In spite of the fact that the number of Japanese speakers in the Kingdom is still low, due mainly to the "King Abdullah Scholarship Program", the number of Saudi students who study in Japan has increased since 2006. Starting with 20 students in 2006, it has increased to 200 in 2008 and to 320 in 2010, finally reaching 500 students in 2012, with students specializing in different majors.

The Kono Initiative

With the beginning of the new century, the Japanese renewed and strengthened their "smart power" diplomatic efforts in the region. These renewed diplomatic efforts were revealed through the visit of the Minister of Foreign Affairs, Mr. Kono Yohei, to the Arab Gulf states on the 8th -10th January 2000. He visited Qatar, the United Arab Emirates, the State of Kuwait, and the Kingdom of Saudi Arabia. This was the first visit by a Japanese Foreign Minister to the Gulf countries since Foreign Minister Nakayama's visit in August 1990, 10 years earlier. The purpose of his visit was to establish multi-layered relations with Gulf countries that extend beyond the import of oil and natural gas. Presenting, "Toward Multi-layered Relations with Gulf Countries (Kono Initiative)," he enunciated that the initiative focused on three areas:

1. Promotion of "dialogue among civilizations with the world of Islam".
2. Water resources development.
3. Promotion of a wide-ranging policy of dialogue.

After announcing this initiative in Qatar on 9th January 2000, and during a television interview, Foreign Minister Kono also discussed these plans during talks with regional leaders, gaining their approval. Moreover, Foreign Minister Kono showed a strong interest in Islam and emphasized the necessity of mutual understanding to explain his stance on attaching the importance of Islam to the initiative. This stance was welcomed by the Gulf countries, which regarded it as very opportune. Under the title of "Outline of Foreign

Minister Kono's Policy Address", the Ministry of Foreign Affairs explained the promotion of "dialogue among civilizations with the world of Islam" on its home page as follows:

> In the course of globalization, the civilization and culture fostered by our respective histories should be cherished. To this end, it is essential that both sides gain a deeper knowledge of each other's civilization and foster understanding and generosity through dialogue with one another. More specifically, Japan will work toward expanding and reinvigorating academic exchanges, intellectual exchanges, and personnel exchanges, among youth (Ministry of Foreign Affairs of Japan, 2011).

The Kono Initiative made it clear that Japan was keen on building a knowledge-based economy by taking part in the development of human resources in several key areas, and that Japan wanted to familiarize societies in the Gulf countries with Japanese concepts through continuous dialogue.

Three months later, Minister Kono funded the "Islam Research Meeting", with the participation of eminent professors specialized in Islamic and Arabic Studies as core members, aiming to enrich Japanese diplomacy through strengthening relations with the Islamic World and Arab countries. That was one year before the United Nations declared 2001 as a "year of dialogue among civilizations", the same year of the September terrorist attacks in the United States.

Despite the fact that Kono had left office in 2001, and the invasion of Iraq took place in 2003, the Ministry of Foreign Affairs of Japan kept holding seminars for "Dialogue Among Civilizations between Japan and Islamic World." The aim was to realize frank and active dialogue amongst scholars and experts from Japan and Islamic countries. It was expected that meaningful discussions would be held to promote mutual understanding between Japan and Islamic countries through seminars. The first seminar entitled "The Islamic World and Japan", was held in Bahrain, 12-13 March 2002, and the second seminar entitled "Peace and Human Development", was held in Tokyo 08 October 2003.

The Ministry of Foreign Affairs of Japan also established "The Japanese-Arab Dialogue Forum" as a part of the initiative, to strengthen the dialogue between Japan and Arab countries. The members of the Forum were influential figures in various fields from Egypt, Saudi Arabia, and Japan. The Japanese members were headed by the late Mr. Hashimoto Ryotaro, former Prime Minister, the Egyptian members were headed by Dr. Ismail Serageldin, Director of the Library of Alexandria, and the Saudi Arabian members were headed by Mr. Khaled Algosaibi, Minister of Planning and Economy. They have all participated in the Forum meetings. The first session was held in Tokyo from 4 to 5 September 2003, the second was held in Alexandria, Egypt from 3 to 4 March 2004, and the third one was held in Riyadh, Saudi Arabia from 8 to 9 January 2005 (Embassy of Japan in Egypt, 2015.).

In 2010, the year of the so-called "Arab Spring", the Ministry of Foreign Affairs of Japan enhanced its activities, holding a series of talks as part of its "Tokyo Seminar" entitled "Future-oriented Dialogue between Japan and Islamic World". Notably, Japanese enterprises, educational and research institutes also embarked on knowledge-based initiatives with Gulf institutes, such as those in UAE and Qatar.

UAE and Qatar

The "Arab Spring" was not Japan's sole motive for reactivating its initiative in the Gulf Area. Also instrumental was the 2011 Tohoku earthquake and tsunami, also known as "Higashi Nihon Daishinsai", which had disastrous consequences on Japan's energy sector. An energy white paper, approved by the Japanese Cabinet in October 2011, stated that the "public confidence in the safety of nuclear power was greatly changed by the disaster and called for a reduction in the reliance on nuclear power". It also omitted a section on nuclear power expansion that was in the previous year's policy review (World Nuclear Association, 2015). The earthquake led Japan to increase oil imports from the Gulf to make up for its loss in nuclear energy.

In the same year, the Institute of Applied Technology in Abu-Dhabi, Cosmo Oil Company of Japan, Abu-Dhabi Oil Company Ltd. (Japan) and Ritsumeikan University in Japan, signed an MOU to establish three-year Japanese language programs for high school students at the Institute of Applied Technology. The purpose of the program was to enable students to join Japanese universities, and also to join summer training at Japanese companies, hence achieving the goal of transferring Japanese technology to Abu-Dhabi (Emarat Alyoum, 2011). The Japanese responsibility in the agreement was to dispatch language instructors, whereas the Institute of Applied Technology was to offer the appropriate atmosphere for teaching. This was the first time that the Japanese private sector participated in the education-centered policies long advocated by the Japanese state.

The culmination of these agreements is a deeper participation in economic and educational sectors in both countries. The Qatari Government focuses on education and the development of human resources as part of this relationship.

The Japanese Government and Japanese private sector are trying to deepen the bilateral relations in all fields, especially in energy, since Qatar has been a major supplier of energy to Japan. In fact, it is one of Japan's largest LNG suppliers, with a share of 18% and total import of about 16 million tons out of 77 million Qatari production of LNG (in 2013).

For these reasons, the Japanese have been working closely with the Supreme Education Council (SEC) to implement the Japanese Educational Experience in Qatar since 2012. The SEC has introduced the Japanese language to some independent high school curricula as an optional second foreign language. Also, graduates from the Japanese Language and Literature Department of Cairo University have been recruited to work as teachers and instructors in these independent high schools. In addition, the SEC has included more than 15 of the top Japanese universities on its scholarship destinations' list for undergraduate and research students. Eight of these selected universities offer academic degrees taught in the English language as part of the program "Global 30", an educational project sponsored by the Japanese Government aimed at nurturing internationally competent individuals by selecting 13 top Japanese universities to offer degree programs for international students wishing to study in Japan, without being hindered by language barriers (Qatar Banks, 2015).

At the level of higher education, Qatar University has signed several MOUs and agreements with prestigious Japanese universities, promoting student and expert exchange, as well as promoting cooperation in the field of scientific research. Another notable development in the academic field is Qatar University's initiative that started in 2014. It has offered the "Marubeni Course in Humanities", which includes lectures on Japanese Studies (language, history, and society) at the Faculty of Arts and Science for undergraduate students. The recruited facultyserving and this program are also from the Egyptian staff (including the author, who is currently the chair-professor in Humanities for this course) at the Japanese Language and Literature Department in Cairo University.

The role that Egyptian individuals in Qatar and other parts of the Gulf are playing as "smart power' mediators in the Japan-Gulf Knowledge Network is obvious. They transfer a part of the knowledge needed, and encourage local talent to succeed in improving their knowledge-based societies, an enterprise which will take years to achieve.

Given that Qatar is quickly diversifying its economy into a knowledge-based one, Mr. Shingo Tsuda, the Ambassador of Japan to Qatar, said in an interview with "Qatar Banks" magazine:

> The impressive quality about Qatar is its ability to maintain equilibrium between an oil-based and a knowledge-based economy in such a way as to help diversifying the country's economy and guarantee a stable and sustainable business environment. In this sense, Qatar and Japan are working constantly and relentlessly to promote and strengthen the ties and to diversify the domains of their bilateral cooperation beyond the excellent existing relations founded on the energy and trade field (Qatar Banks, 2015: 12-15).

Furthermore, many Qatari technicians and administrators have been invited to join several Japanese companies' training programs. Aimed at creating an effective framework for fostering a qualified young generation, such programs also help Qatar develop industries capable of implementing change in the future. In Qatar, transnational experts and Japanese soft power are playing a great role in guaranteeing a stable and sustainable business environment for safe energy supplies.

Conclusion

As the Japanese approach to the Gulf countries aims at developing transnational knowledge network by relying on experts, smart power, endowments, training programs and educational projects, the soft power of Japanese culture has also made a commitment to regional security.

On 29 October 2013, a Seminar on "Construction of Partnership for A New Era with the Islamic World" held in Doha, Qatar, by the Ministry of Foreign Affairs of Japan in cooperation with Brookings Doha Center, offered a roundtable discussion on "Gulf Security"; it promoted defense cooperation through the soft power of Japanese culture. Such initiatives will help Japan fulfill a political, yet exceptional role in the Middle East (Ministry of Foreign Affairs of Japan, 2015).

What needs to be taken into consideration, however, is Japan's current situation. Factors, such as the change in the balance of power sparked by China's foreign expansion and US' gradual troop reduction in the Middle East, as well as economic conditions and energy supply problems following the earthquake disaster, need further consideration. China is likely to compete with Japan over access to the energy supply sources and markets in the Gulf countries, and it will be necessary to employ such systems for the purpose of international trust building and the prevention of any 'free-riders.' Each of these factors will have an impact on Japan's defense and diplomatic policies and, by extension, the Japanese approach to the Gulf.

For Japan, the importance of its relationship with the Gulf countries will surely trigger the expansion and deepening of these ties in the foreseeable future. However, the immediate deployment of Japanese Maritime Self-Defense Forces to the Indian Ocean and the Gulf Region is difficult. It would be an unconstitutional act under the ninth article of the current constitution of Japan. As such, Japan's role in maintaining regional security will probably depend on the licensing of technology to India and the Gulf countries, as well as the construction of collaborative systems devoted to the training and support of highly proficient personnel.

This article explains how Japan is committed to strengthening a knowledge-based economy in the Gulf countries through its soft power, and the transnational individuals it utilizes for its "smart power" strategy. The response of the Gulf countries towards Japan's efforts, however, still seems lackluster. For example, 20 years after its founding, the Japanese language program at King Saud University with the exception of two young teaching assistants has not produced any native faculty members. In addition, the agreement between the Applied Technology Institute of Abu Dhabi and Japanese enterprises and universities, while very practical, is still limited to one institute. In Qatar's case, in spite of Japan's special attention to diversifying their relations and Qatar's eagerness to improving this bilateral relationship, the attainment of a knowledge-based society still faces problems. Many of the skilled individuals working at the technical and educational institutes, which benefit from the cooperation with Japan, are not Qataris. In the final analysis, locals will not be able to participate in the knowledge-based society, nor help their country after the end of their short-term contracts. In fact, these short-term contracts are a hindrance to the training of a sufficient number of national experts. Although the nationalization policy in the Gulf countries aims to reduce dependence on transnational experts, this process is expected to take even longer to achieve when it comes to producing experts in Japanese studies.

References

Action Forex (2004) *1973 Oil Crisis*. Available at: http://tinyurl.com/jfkqq4p

Dowty, A. (2000) Japan and the Middle East: Signs of Change? MERIA 4(4): 67-76.

Emarat Alyoum (2011). Available at: http://tinyurl.com/3d4yq7x

Embassy of Japan in Egypt (2015) *Japan-Egypt Relations*. Available at: http://tinyurl.com/j5k454e

Embassy of Japan in Saudi Arabia (2015) *How to learn Japanese in KSA*. Available at: http://www.ksa.emb-japan.go.jp/topics/110709.htm

Hosaka, S. (2015) "sanyu koku denai ejiputo ga chuutou de juuyoushi sareru riyu toha?" [what is the reason of emphasizing Egypt which is not an oil producing country?]. *Mynavi News*. Available at: http://news.mynavi.jp/column/MiddleEast/004/

Ishiguro, H. (2014) "wangan oyobi indoyou chiiki no anzen hoshou to Nippon" [Japan's Commitment for the Regional Security in the Indian Ocean and the Gulf], *chuto rebiyu-[Middle East Review]*, 1: 76-91.

The Japan Foundation (2015) *About Us*. Available at: https://www.jpf.go.jp/e/about/index.html

Ministry of Foreign Affairs of Japan (2001) *The visit to the Gulf Countries by Minister for Foreign Affairs Yohei Kono (Overview and evaluation)* Available at: http://www.mofa.go.jp/region/europe/russia/fmv0101/overview.html

Ministry of Foreign Affairs of Japan (2015) *Security Roundtable for Intellectual Exchanges between Japan and the Gulf Countries*. Available at: http://www.mofa.go.jp/mofaj/files/000018297.pdf

Nakanishi, H. (2011) Wangan Senso to Nihon Gaiko [Gulf War and Japanese Diplomacy], *Nippon.com*. Available at: http://www.nippon.com/en/features/c00202/

Qatar Banks (2015): 12-15.

Sirkhabi, R. (2015) History of Oil: The First Oil-shock. *GeoExPro* 12(3): 92-96.

World Nuclear Association. Fukushima Accident (2015) Available at: http://tinyurl.com/gt2crta

7

Japanese Multinationals Breaking with their Business Groups: Opportunities for Global Organizations

Camilla Nellemann

Introduction

Japan was until 2010 the world's second largest national economy and remains the third-largest, and it has long been a significant energy-importer from the Gulf States. In light of Japan's economic, political and technological weight in the international economic and political system, Japanese multinational corporations are potentially important partners in transnational knowledge relations for the Gulf. My research shows that the pursuit of foreign partnerships by J-MNCs represents an opportunity for outside firms – as in the Gulf – to engage with J-MNCs and benefit from potential knowledge spillovers both in technology and business practices, where Japan has unique characteristics. There is a strong tradition for collaboration among Japanese companies. Interlinked firms "pervade much of the Japanese economy" (Gerlach, 1992: xiv). Japanese network organization was seen as a driver for the remarkable competitiveness of J-MNCs in the 1980s, yet many argue today that they advanced in spite of it (Lincoln and Gerlach, 2004). As the world is turning into a 'global village' with economic activity becoming more interdependent (Dunning, 1997), successful integration into global value chains is crucial for MNCs (Pudelko, 2005) Many J-MNCs have indeed found that limiting their partnerships to domestic ones no longer suffices (Teramoto and Benton, 2005). Therefore, two cases of J-MNC transnational partnering are discussed and analyzed in this chapter: a joint venture between Mitsubishi Heavy Industries and Vestas in Denmark, and a joint venture between SBI Pharmaceuticals and Dawani in Bahrain.

J-MNCs breaking with their Japanese business groups to better integrate into the 'global village' represents an opportunity for Gulf and other foreign firms and organizations at large capable of offering assets that can help to advance the business of J-MNCs. There is evidence of foreign partners benefitting from knowledge spillovers in partnerships with J-MNCs (Kantei, 2013); Toyota Motor Corporation, for example, supports the Saudi-Japanese Automobile High Institute in Saudi Arabia which up until now has trained more than a thousand automotive engineers. Most of the graduates become employed by Saudi distributors and dealers of Japanese automobiles. The institute helps to advance the level of automotive technology in Saudi Arabia and contributes to the Saudi-Japan Industrial Cooperation (Toyota, 2015). This framework for cooperation was established by the governments of Saudi Arabia and Japan in 2007. It features three pillars: assistance for Japanese firms looking to invest in Saudi Arabia; assistance in human resource development in Saudi Arabia; and assistance in formulating small and medium enterprises development policies (Japan-Saudi Arabia Industrial Cooperation Task Force, 2012: 1). Another example is the sponsorship by Japan Oil Development Company of the Petroleum Institute (PI) in Abu Dhabi. This institution specializing in engineering education and energy industry research was established in 2001. Currently, PI has around 2000 undergraduate and graduate students, some 200 faculty, and has developed into a leading teaching and research institution in the Middle East. Besides Japan Oil Development Company, (JODCO) PI's sponsors include the Abu Dhabi National Oil Company, BP, Shell, and Total. The institution opens the PI Research Center (PIRC) in 2015 (PI, 2015). Given the low energy self-sufficiency rate of Japan, the country has long depended greatly on energy imports from the Middle East region (Uchida, 2013). In 2013, for example, 31.4 percent of Japan's oil import came from Saudi Arabia, while 21.1 percent from the UAE, Qatar contributed with 12.5 per cent, Kuwait with 7.3 per cent, Iran with 5.9 percent, and Oman with 2.3 percent (Burret, 2014). JODCO is a wholly-owned subsidiary of the Tokyo-based J-MNC INPEX Corporation. It has been engaged in oil production in Abu Dhabi for more than 40 years through JODCO. Currently, the company produces oil from the Nasr Oil Field offshore Abu Dhabi jointly with the other sponsors of PI (INPEX, 2015). In 2014, PI completed a Solar Car developed in cooperation with INPEX and Japan's Tokai University. The unveiling ceremony was attended by H.E. Abdulla Nasser Al Suweidi, Director General of Abu Dhabi National Oil Company and chairman of the PI Governing Board. INPEX provided assistance in developing the relationship between the two educational institutions. The project draws on the technology and expertise of Tokai University's Solar Car Team having recorded excellent results in international solar car races, and is designed to transfer project management skills to the students of PI (INPEX, 2014). In this way, INPEX combines its business activities in Abu Dhabi with social contributions (INPEX, 2015).

The business environment facing J-MNCs has undergone significant change since what came to be known as the 'Japanese management model' developed in the post-war era. Consequently, many authors point to the fact that this model is in a state of flux (Kono and Clegg, 2001; Abe, 2010). A key feature of the model is the embeddedness of J-MNCs in business groups. In this chapter, I analyze the motivations of J-MNCs to break with their business groups to pursue growth and enhance their competitiveness through transnational partnerships. In doing so, I first review the literature on corporate grouping in Japan to determine how this management practice has evolved over time. I then outline and discuss two case studies, the joint venture between Mitsubishi Heavy Industries and Vestas in Denmark, and the joint venture between SBI Pharmaceuticals and Dawani in Bahrain.

My case studies show that J-MNCs join forces with foreign peers to acquire the various assets they need to grow and compete globally. I thus confirm the hypothesis of Aoki (1994) and Dore (1994), implying that a shared sense of 'Japaneseness' in the Japanese corporate world, which is well documented by many authors (Aoki, 1994; Anchordoguy, 2005; Dore, 1994; Porter, Sakakibara, and Takeuchi, 2000), is likely to diminish amid increasing pressure from global competition. I conclude with the argument that the pursuit of foreign partnerships by J-MNCs represents an opportunity for global firms to engage with J-MNCs and benefit from potential knowledge spillovers.

Following this introduction, I provide a review of the literature on Japanese business groups. I then explain my methodology before presenting my case studies. In the subsequent discussion section, I compare the cases with the literature review to identify the motivations of J-MNCs for developing transnational partnerships. Lastly, I conclude and suggest future research.

Literature Review

The Main Bank System and Corporate Grouping

A unique feature of J-MNCs is their embeddedness in corporate groups. Following the Second World War (WW II), the main bank system developed in Japan to circumvent the problem of scarce capital and offer firms the opportunity to raise funds at a low cost. Aoki and Patrick (1994: xxi) describe the main bank system as "a more or less informal set of regular practices, institutional arrangements and behavior that constitute a system of corporate finance and governance". It is an implicit organizational form (Hoshi, 1994); there is no formal or legal definition of a main bank. Under the main bank system, J-MNCs obtain their largest share of borrowings from one bank. The main bank is the largest but not sole lender to the J-MNC. It is also the J-MNC's largest stockholder and acts as a monitor. The main bank is in a unique position to do so as it operates the J-MNCs major payment settlement accounts. The main bank, furthermore, has the right to intervene in the management of the J-MNC when things go wrong (Aoki, Patrick, and Sheard, 1994). The main bank system allows the J-MNC to enact strategic decision with no regard for the interests of individual stockholders (Aoki, 1988). It also allows coordinated refinancing (Hoshi, 1994), and, therefore, can reduce the costs of financial distress (Sheard, 1994; Hoshi, Kashyap, and Scharfstein, 1991). Finally, the system helps to reduce the information and incentive problems inherent in arms-length financial contracts (Hoshi, 1994).

The main bank system gave rise to two sorts of corporate groupings in Japan (Aoki, 1988). One kind is the horizontal *keiretsu,* where unrelated firms are united through a group of financial institutions, including their main bank. The other type is the vertical *keiretsu,* where related firms are organized under a principal manufacturer, which often is a member of a horizontal *keiretsu* as well (Hoshi, 1994). Cross-shareholding (*kabushiki mochiai*) is also closely related to corporate grouping (Hoshi, 1994). Most J-MNCs have extensive cross-shareholding with partners somewhat related in their value chain, i.e., banks, insurance companies, suppliers, customers, trading companies, and affiliated firms (Sheard, 1994). These corporate shareholders hold each other's shares as 'stable shareholders' (*antei kabunushi*), that is they refrain from selling the shares to speculators and third parties who are unsympathetic to incumbent management. In case selling is necessary they agree to notify the firm of their intention to sell. Hence, they cannot claim the property rights that usually are associated with shareholding and they are excluded from exercising corporate control. Trust is crucial to sustaining cross-shareholding because the compensation is a reciprocal exchange of property rights (Sheard, 1994). This practice intends to insure against financial failure and external take-over (Aoki, 1994). It is considered a competitive advantage of J-MNCs "enabling management to take a long-term view of investment and decision-making without being subject to short-term stock market pressures" (Sheard, 1994: 311).

The Horizontal Keiretsu

The members of a horizontal *keiretsu* are the large firms belonging to Japan's economic, political and social elite (Gerlach, 1992). They are united through six kinds of linkages where one is cross-shareholding. Another is the monthly presidents' council meeting (*shachōkai*), typically including four financial institutions, which are the major debt holders of the J-MNCs. Third, their main bank often sends an employee to each J-MNC who becomes a permanent employee of the firm and serves as a board member for at least 10 years. This employee is an important source of information for the main bank enabling efficient monitoring of the J-MNC. Fourth, the main bank may, if necessary, send its own employees to act as directors of the J-MNC. Transactions in intermediate product markets are the fifth linkage. Lastly, J-MNCs in a horizontal *keiretsu* may undertake joint projects, e.g., collaborate to set up a company in order to start a new line of business (Hoshi, 1994). Due to the main bank's central position in the presidential council and the horizontal *keiretsu* it can "act as *de facto* owners of *keiretsu* member firms" (Oh, 2004: 66).

The Vertical Keiretsu

The members of a vertical *keiretsu* are smaller and less prestigious firms compared to those in a horizontal *keiretsu.* These firms operate within the same industry and under the control of a J-MNC (the principal manufacturer). There are three categories of vertical *keiretsu:* the production *keiretsu* comprising hierarchies of primary, secondary and tertiary-level subcontractors supplying the J-MNC; the distribution *keiretsu* which constitutes the vertical marketing system of the J-MNC; and the capital *keiretsu* consisting of affiliated companies depending on capital from the J-MNC (Gerlach, 1992).

The J-MNC tends to dominate the subcontractors because the former utilize a large amount of resources in selecting the latter, and technological know-how is transferred from the J-MNC to the subcontractors. The ability to dominate depends on the amount of capital, not the supply of material (Oh, 2004). Whittaker (1994), however, emphasizes strong interdependence within the vertical *keiretsu* where the J-MNC greatly depends on its subcontractors due to their mutually beneficial relation. In fact, subcontractors may at some point become independent companies. Not least since pressure from the J-MNC to reduce costs, raise their technical level, rationalize production, and participate in design and development help them to enhance their capabilities and eventually launch their own products to a wider audience.

In figure one, below, I have illustrated the two kinds of corporate grouping. The box to the right shows the principle of the horizontal *keiretsu* with unrelated J-MNCs united with each other and financial institutions. The boxes to the left show how each member of the horizontal *keiretsu* controls its own vertical *keiretsu* including a production *keiretsu* (P), a distribution *keiretsu* (D) and a capital *keiretsu* (C).

Figure 1: Post-WW II Corporate Grouping in Japan

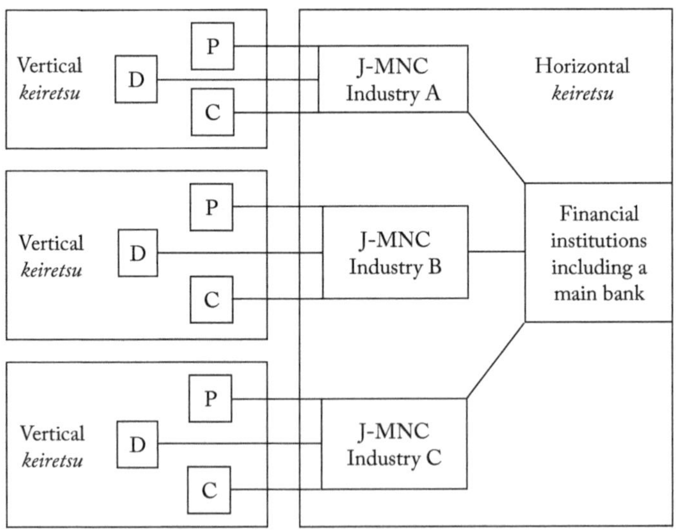

Source: the author

The number of firms in the horizontal *keiretsu* is not limited to three and some vertical *keiretsu* may not include all three kinds (P, D, and C) such as indicated in my figure above. The reality is that it is difficult to map out these webs exactly as firms may hesitate to inform about them and the business groups lose some and gain other members over time (Oh 2004). They are dynamic and adapt to changing circumstances (Lincoln and Gerlach, 2004).

Trends in Corporate Grouping

A 1987 survey indicated that nearly all Japanese companies had a main bank, and, in turn, every Japanese bank served as a main bank to a number of firms. A good relationship with one's main bank used to be essential for corporate success. This changed somewhat during the 1980s when firms turned to issuance of market securities inside and outside of Japan (Aoki et al., 1994). Even so, the main bank system is alive and well. In a 2003 survey among the Nikkei 500 Stock Average companies, 94 percent mention one or more particular banks as their 'main bank', and a fifth of these firms had increased their main bank borrowings (Abegglen, 2005). Accordingly, Witt (2006) concludes that the Japanese financial system continues to be bank-led, albeit corporate funding from banks modestly declined from 74.2 percent in 1989 to 63.8 percent in 2002. At the same time, however, J-MNCs increasingly raise funds on the stock market. They have attracted foreign investors (Teramoto and Benton, 2005), who previously were held at bay (Porter et al., 2000; Anchordoguy, 2005). Given these trends, the main bank system is likely to continue but will play a lessened role in the future (Abegglen, 2005).

Scholars (see Porter et al., 2000) attempting to explain the Japanese economic crisis having lingered on since the 1990s, point to patient capital entailed by cross-shareholding as a key factor. Still, *Keiretsu* organization continues to be widespread. According to Witt (2006), 52.8 percent of member firms anticipate that their corporate group will maintain the present state or deepen while only 18.4 percent expect it to dissolve. On the other hand, there are some indications that *keiretsu* ties are weakening. For instance, intra-group purchasing decreased from 7.75 percent in 1992 to 6.44 percent in 1999. The fact that the current trend in *keiretsu* organization is hard to determine can be illustrated by the example of Nissan which cut off a number of its *keiretsu* ties in order to turn around the near-bankrupt corporation, while, in contrast, Toyota has deepened the relationship with its suppliers even further to prevent them from collaborating with Toyota's rivals.

Overall, cross-shareholding and stable shareholding have nonetheless seen a marked decline since the burst of the bubble economy in 1991 when firms and banks faced a liquidity crisis, and were forced liquidate some of their shares (Witt, 2006). Clearly, the reciprocity acting as glue between *keiretsu* firms has been hard to sustain during times of financial pain. Consequently, the degree of foreign ownership has increased (Lincoln and Gerlach, 2004). These trends are in line with Dore's (1994: 390) prediction: "(...) Japan's competitors (…) can look forward to the erosion of togetherness (…) as Japanese society becomes progressively internationalized". Similarly, in a discussion of how the Japanese economy might evolve in the future, Aoki (1994: 36) asserted the following: "It is not effective for the Japanese to defend the rationality inherent in the J-system, since the closed nature of its institutional framework needs to be modified to enable it to become enmeshed in the global system in a more consistent way".

Nowadays, J-MNCs clearly see the benefits of collaborating with foreign peers. Nissan, for example, formed an alliance with French Renault in 1999 to avoid bankruptcy, whereas Seiyu formed an alliance with American Wal-Mart in 2002 to benefit from the latter's management expertise (Teramoto and Benton, 2005). Foreign peers are becoming part of the *keiretsu* (Lincoln and Gerlach, 2004). Since the 1980s, J-MNCs' preferred mode of international expansion has been the strategic alliance where they are able to utilize their extensive experience with domestic alliances (Westney, 2006).

Methodology

I pursue the case study methodology to analyze the motivations of J-MNCs to develop cross-international partnerships. A case study is well suited for 'How'- and 'Why'-research questions (Yin, 2009). It allows researchers to study a phenomenon within its context based on several kinds of data (Baxter and Jack, 2008). The case study is "an intensive study of a single unit for the purpose of understanding a larger class of (similar) units" (Gerring, 2004: 342). In my research, each of the cases represents a single unit that helps me to develop a general argument about J-MNCs' inter-firm collaboration.

Case studies can be designed in various ways (Piekkari and Welch, 2011). I employ the Critical Discourse Analysis (CDA) approach described by Ahonen, Tienari, and Vaara (2011). CDA is interdisciplinary and questions taken-for-granted assumptions to provide alternative ways to understand internationalization and globalization as well as their causes and implications. Essentially, CDA attempts to distinguish discourses through which social and organizational reality is constructed. In this process, it is relevant to analyze issues related to power and ideology reflected in social relations; contextuality including historicity; and the intertextuality and interdiscursivity which is evident when comparing different data (Ahonen et al., 2011).

Buck (2011) emphasizes the importance of carefully selecting the case study with consideration to theory rather than convenience. I selected these cases because they offer insights about J-MNCs' transnational partnerships. My considerations regarding case selection correspond to the 'theoretical sampling' strategy conceptualized by Fletcher and Plakoyiannaki (2011) implying to research cases which may replicate or extend the emergent theory. I decided to conduct more than one case study because Eisenhardt (1989) explains that such an approach enables cross-case search for patterns, which helps to prevent jumping to premature conclusions based on limited data. Fletcher and Plakoyiannaki (2011: 185) agree that the multiple case design offers "greater opportunity for generalizability and external validity". It enables the researcher to analyze differences both within and between cases (Yin, 2009).

My data consists of a semi-structured interview of two managers at Toshiba conducted in Kawasaki, Japan, on April 14, 2015. The interview lasted for one hour, the language spoken was Japanese and I took notes during the interview. It illuminates how the interviewees perceive their reality to better understand their behavior, in line with the constructivist approach assuming that truth is relative (Baxter and Jack, 2008). In addition, I utilize a variety of textual data, i.e., public reports of companies, agencies, and ministries and relevant media reports. As the collection of textual data must be guided by the research question (Ahonen et al., 2011), I evaluated how the particular data can help me to answer why the J-MNCs I examine pursued international partnerships. Mills and Mills (2011) stress that textual data may contain information that interviewees either cannot or will not say during an interview. Thus, combining interviews with textual data is likely to bring more aspects to light.

I report each case study as a narrative describing the particular inter-firm collaboration. Mir (2011) argues that one of the most important tasks of a researcher pursuing the case study approach is to 'narrativize' the collected data, thus produce a coherent story. In this process, the researcher decides which data to include and which to neglect.

Analysis

Case I: Mitsubishi Heavy Industries in Denmark

A severe nuclear accident broke out in Fukushima, Japan, in 2011. The accident occurred in conjunction with the Great East Japan Earthquake and subsequent tsunami and was exacerbated by an inadequate organizational response. The official investigation of the accident determined that TEPCO, the operator of the disaster-struck Fukushima Dai Ichi Nuclear Power Plant, had succeeded in neglecting safety issues throughout many years due to regulatory capture (NAIIC, 2012). Consequently, a new and independent regulatory body has been established in Japan to enhance the nation's nuclear safety and to the reclaim the public's trust in the government's administration of nuclear power (NRA, 2015). A measure which is much needed considering post-Fukushima public opinion surveys indicating considerable opposition to nuclear power among the Japanese population (Aldrich, 2014; Hobson, 2014; Kingston, 2014). The Fukushima accident highlighted the importance of energy diversification. Since the accident, the Japanese state has intensified its efforts to promote the use of renewables (METI, 2014).

MHI was founded as a shipbuilding company in Japan in 1884 called Nagasaki Shipyard & Machinery Works. In 1934, the company had developed into Japan's largest private enterprise. It was re-launched as a manufacturer of ships, heavy machinery, airplanes, and railway carriages under the name Mitsubishi Heavy Industries. Today, MHI is involved with engineering, manufacturing and the sale of ships, power systems, environmental improvement equipment, industrial machinery, aircraft, space systems, and air conditioners, among other activities. It covers the following business segments: Energy & Environment; Commercial Aviation & Transportation Systems; Integrated Defense & Space Systems; and Machinery, Equipment & Infrastructure. In fiscal 2014, the Energy & Environment segment accounted for 37 percent of MHI's total net sales (MHI Report, 2014). MHI has engaged in the research and development of wind energy systems since 1980 and possesses a number of complementary skills (MHI-Vestas, 2015). MHI delivered Japan's first wind turbine for commercial use in 1982 and had by 2013 supplied more than 4,200 units worldwide with a combined output of around 4,400 MW. Before the partnership with Vestas, MHI was undertaking the development of a 7 MW wind turbine called the Sea Angel applying the world's first digitally controlled hydraulic drivetrain. MHI is a member of the consortium behind the Fukushima FORWARD project developing a floating offshore wind farm (Vestas, 2013). MHI belongs to the Mitsubishi *keiretsu* (Oh, 2004).

Along with a changing sentiment over nuclear energy in Japan, collaboration between MHI and Vestas on offshore wind power came into being. The dialogue between the two firms began in 2012. It was, apparently, initiated by the Japanese government, which encouraged the management of MHI to establish contact with Vestas. Japan had the financial capacity to develop its offshore wind industry, however, it lacked the technology. The Japanese government and MHI opted for a partnership with Vestas due to the technological position of Vestas and Japan's positive relations with Denmark. At the same time, it was clear that Vestas was in financial difficulty (Energy Watch, 2013). An analyst remarked that Vestas lagged behind its competitor Siemens in the offshore market and would benefit from having a 'big brother', i.e., MHI (Reuters, 2013). Vestas was founded in 1898 as a blacksmith business in Denmark. In 1945, the company was relaunched under the name Vestas. It began manufacturing kitchen appliances and agricultural equipment and become an exporter in the 1950s. Vestas began experimenting with wind energy systems in the 1970s and eventually developed into a streamlined windmill manufacturer (Vestas, 2015). Vestas is a pioneer of offshore wind energy since 1990. The company installed its first offshore wind turbine in 1991, and its first offshore wind farm in 1995 (MHI-Vestas, 2015). By the time of entering into a partnership with MHI, Vestas enjoyed the position as the world leader in wind energy measured by cumulative installed capacity (57,000 MW). The company had provided wind energy to 73 countries through the installation of more than 50,000 wind turbines (Vestas, 2015).

The joint venture between MHI and Vestas was announced in the autumn of 2013 (Børsen, 2013), and formally established on April 1, 2014. The entity adopting the name MHI-Vestas Offshore Wind is headquartered in Aarhus, Denmark (Vestas, 2014). Vestas and MHI have each transferred 170 employees to the joint venture, which they own 50/50 (MHI-Vestas, 2015). However, MHI has an option to increase its ownership to 51 percent in 2016 (Vestas, 2013; MHIE, 2015). The board consists of an equal number of members from MHI and Vestas and is chaired jointly by the Chairman of MHI Europe,

Mr. Masafumi Wani, and the CEO of Vestas, Mr. Anders Runevad. The management also consists of an equal number of former MHI and former Vestas employees (MHI-Vestas, 2015). MHI-Vestas Offshore Wind initially commenced business with Vestas' V112-3.0 MW wind turbine and began offering the V164-8.0 MW wind turbine from 2015 (MHIE, 2015). The two partners complement each other as follows:

> MHI's strong financial position and a long-standing presence in global power markets and related technologies are a clear strength to MHI Vestas Offshore Wind™. Vestas' comprehensive technological capabilities and development of the V164-8.0 MW®, combined with a long-standing track record within the overall wind power market, and the offshore wind segment, is a key contribution to MHI Vestas Offshore Wind™. Leveraging on these respective strengths of each organization, MHI Vestas Offshore Wind™ will benefit from significant synergies, and thereby be in a very strong position to win an expanding share of the offshore wind turbine market and to lead the growth (MHI-Vestas, 2015).

MHI-Vestas Offshore Wind is responsible for the design, further development, procurement, manufacturing, installation, commissioning and service related to the V164-8.0 MW turbine as well as marketing, sales and after-sales service related to offshore wind (MHI-Vestas, 2015). MHI continues to develop the DDT technology which is expected to be mature enough to be integrated into MHI-Vestas' product portfolio later this decade. None of the partners are allowed to enter into the offshore wind market outside of their joint venture (MHIE, 2015).

In July of 2014, MHI-Vestas gained an order from Skovgaard Invest ApS and Energicenter Nord of four V164-8.0 MW offshore wind turbines for the Velling Mærsk project in Denmark (MHI-Vestas, 2014). By now, it has obtained several orders in Northern Europe (MHI-Vestas, 2015; Dong Energy, 2015). In March of 2015, MHI-Vestas announced that the company is ahead of its plans and will double the number of employees in the near future (Børsen, 2015). The partnership with Vestas looks to fulfill MHI's ambition of expanding its wind power business: "Through the new joint venture we have established with Vestas Wind Systems A/S, we will strive to develop the offshore wind turbine business on a global scale" (MHI Report, 2014: 27).

The Mitsubishi *keiretsu* is known to remain a strong, loyal network even today (Mitsubishi, 2015, personal communication). In 2004, for instance, the Mitsubishi *keiretsu* bailed out the poorly performing Mitsubishi Motors (The Economist, 2015). Recently, however, MHI is pursuing a new business structure involving less vertical relationships to better seize opportunities and enhance growth. This is done by merging some of its businesses, divesting the less value-adding activities, and establishing partnerships with other firms (MHI, 2010). J-MNCs have long been criticized for their hesitation to terminate unprofitable activities and failure to allocate resources to the most promising lines of business instead of spreading them across the entire portfolio (Westney, 2006).

Case II: SBI Pharmaceuticals in Bahrain

For quite some time, the Gulf countries have been eagerly trying to attract foreign MNCs in attempting to diversify their economies (SWIFI, 2013). At the same time, the business opportunities in the Gulf appeal to many J-MNCs. As Toshiba, a Japanese conglomerate, so well puts it: "we hope to install the elevators in all their tall buildings!" (Toshiba, 2015, personal communication). From the Japanese perspective, agriculture, energy and health care are especially fruitful areas for the cultivation of partnerships with Gulf firms (Kantei, 2015).

Bahrain aims to transform its capital into a regional biotech hub by attracting pharmaceutical companies. In October of 2012 SBI Pharmaceuticals established a representative office in Bahrain (SWIFI, 2013). This Tokyo-based J-MNC, previously known as SBI ALApromo, is a joint venture between SBI Holding and COSMO OIL founded in 2008. SBI Pharmaceuticals researches, develops and manufactures health products using the 5-Aminolevulinic Acid (5-ALA). The 5-ALA is found in animals and plants and is an ingredient for hemoglobin in blood and chlorophyll. SBI Pharmaceuticals mass produces 5-ALA utilizing technology developed by COSMO OIL. It covers three business areas. In the medicine business, the products of SBI Pharmaceuticals helps to diagnose and eventually treat cancer. In the cosmetics business, the company provides anti-aging skin care products, and in the health food business, it supplies dietary supplements addressing lifestyle diseases and anemia (SBI Pharmaceuticals, 2015).

In March 2013, SBI Pharmaceuticals increased its operations in Bahrain with the entrance into a Memorandum of Understanding (MoU) with the Manama-based Mumtalakat (SBI Holdings, 2015). Mumtalakat is an investment organization owned by the government of Bahrain. It manages the sovereign wealth fund for the Kingdom of Bahrain. It's fund's deposit mainly come from the surplus revenues from the development of gas and oil reserves in the country. Mumtalakat is managed by a nine-member board of directors featuring public and private sector experts. Its investment portfolio is diversified and concentrated on local and regional firms, which substantially contribute to the Bahraini economy (Investopedia, 2015). Mumtalakat invests with the purpose of securing the future wealth of the Bahraini people. It currently has 38 equity positions in the seven sectors Logistics, Consumer & Healthcare, General Services, Financial Services, Industrial Manufacturing & Services, Real Estate & Tourism, Telecommunication, and Technology & Media. Mumtalakat Holdings not only invests in companies but actively contributes to their management (Mumtalakat, 2015). The portfolio of Mumtalakat is worth USD 7.1 billion (SBI Holdings, 2015a).

The purpose of the partnership between SBI Pharmaceuticals and Mumtalakat is to explore the potential for the establishment of a pharmaceutical lab in Bahrain (SWIFI,

2013). At the time, SBI Pharmaceuticals was preparing to start a clinical study on diabetes by using 5-ALA at the Diabetes Department of the Bahrain Defense Force Royal Medical Service Hospital, the Arabian Gulf University Hospital, and the King Hamad University Hospital, among other locations in Bahrain. The goal of SBI Pharmaceuticals was to work with Mumtalakat including its network to establish a manufacturing base in Bahrain and expand SBI Pharmaceuticals' global sales from Bahrain (SBI Holdings, 2013a).

On the same day as signing the MoU with Mumtalakat, SBI Pharmaceuticals also established a joint venture with the Manama-based Dawani Group Holding (Dawani) (SBI Holdings, 2013b). Dawani was created in 2005 to manage the group's business activities mainly in manufacturing, trading, retailing of paints and chemicals, real estate, distribution of food through its own supermarket chain, retailing of garments and education. There are eight companies under the Dawani umbrella, namely Deeko Bahrain, Dawani Properties, Tominna Trading, Collection, Ermenegildo Zegna, Dana Tool Room, Dawanco Industries and Dawanco. These companies have a loyal customer base as well as a good reputation in Bahrain (Dawani Group Holding, 2015). Their joint venture is situated in the North Sitra Industrial Area in Bahrain. It is owned 50/50 by SBI Pharmaceuticals and Dawani, and each company has dispatched a director (SBI Holdings, 2015b). By establishing a joint venture with Dawani, SBI Pharmaceuticals foremost obtains marketing channels in the Middle East:

> The new joint venture company will take advantage of the know-how of both companies, SBI Pharmaceuticals, and Dawani Group Holding, and discuss specific details of marketing activities for 5-ALA products in Bahrain. Our new partner, Dawani Group Holding, is a major distributor and manufacturer that has import companies and companies operating shopping malls in Bahrain under its umbrella and has know-how in the retail business in the Middle East. SBI Pharmaceuticals will make every effort to expand sales of 5-ALA products in Bahrain and in other Middle East regions, through the joint venture company (SBI Holdings, 2013b).

Discussion

As the term 'globalization' proliferated in the 1980s, Contractor and Lorange (1988) argued that managers now need to consider competition and cooperation as equally valid parts of the corporate strategy. The J-MNCs, which traditionally have managed both, emerged as leaders and exemplars in the context of international alliances (Westney, 2006). Dunning (1997) offers four main reasons to explain collaboration across firms: 1) to acquire new product or process technologies; 2) to spread the risk of high capital outlays, or reduce the time of product development; 3) to capture the economies of synergy or scale; and 4) to gain access to new markets or distribution channels.

In case one we saw the Japanese government and MHI exercising agility when facing a changing institutional context following the Fukushima accident. As they became aware of the need to expand the Japanese offshore wind industry, and the associated need to enhance the technological capacity of Japan in this area, they actively pursued the acquisition of these capabilities in the global wind industry. While all four motivations proposed by Dunning (1997) are at play, in this case, the first motivation is particularly relevant. Case two showed how SBI Pharmaceuticals first established its presence in Bahrain, then pursued local partnerships to exploit the business opportunities in the Middle East region. It is especially Dunning's (1997) fourth motivation that applies in this case. Thus, in addition to the motives for transnational collaboration shown in my literature review, i.e., access to capital in the case of the Nissan-Renault alliance, and access to know-how in the Seiyu-Wal-Mart collaboration, the MHI-Vestas partnership and SBI Pharmaceuticals' partnerships with Dawani and Mumtalakat indicate motives such as access to technology and distribution channels. My case studies confirm past scholarship identifying an increasing integration of J-MNCs into global value chains.

Concluding Remarks

In this chapter, I described a trend among J-MNCs to break with their business groups and analyzed why they develop partnerships with foreign firms. I identify the following motives: 1) access to new markets; 2) access to technology; 3) access to know-how, 4) and access to capital. According to my findings, J-MNCs break with their business group when the assets they need to grow and be globally competitive are not available within these domestic networks. The trend represents an opportunity for global organizations capable of contributing to the goals of J-MNCs.

The insights put forward in this chapter help to fill a gap in the literature identified by Witt (2006) on the state of Japanese business groups. My findings shed some light on the topic but more research is needed to develop a clearer picture.

Acknowledgement

My research is funded by the Japanese Ministry of Education, Culture, Science and Technology (MEXT) and the Scandinavia-Japan Sasakawa Foundation. I thank the faculty at Rikkyo University, College of Business; Prof. Pasi Ahonen, Essex Business School; Prof. Michael Witt, INSEAD; Mr. William Lim, Vestas Asia Pacific & China; Ms. Sajida H. Shoff, Almont Group; and Dr. Moharem El Gihani, Cyclofluidic, for their valuable comments.

References

Abe, M. (2010) Introduction: Japanese Management in the 21st Century. In: Haghirian. P. & Abe, M. (eds.) *Innovation and Change in Japanese Management*, London: Palgrave Macmillan.

Abegglen, J.C. (2005) A Perfect Financial Storm. In: Pudelko, M. & Haak, R. (eds.) *Japanese Management: The Search for a New Balance between Continuity and Change*, London: Palgrave Macmillan.

Aldrich, D.P. (2014) Revisiting the limits of flexible and adaptive institutions: The Japanese government's role in nuclear power plant siting over the post-war period. In: Kingston. J. (ed.) *Critical issues in contemporary Japan*, New York: Routledge.

Anchordoguy, M. (2005) *Reprogramming Japan*, Ithaca, New York: Cornell University Press.

Aoki, M. (1988) *Information, Incentives, and Bargaining in the Japanese Economy*, Cambridge: Cambridge University Press.

Aoki, M. (1994) The Japanese Firm as a System of Attributes: A Survey and Research Agenda. In: Aoki, M. & Dore, R. (eds.) *The Japanese firm: The Sources of Competitive Strength*, New York: Oxford University Press.

Aoki, M. & Patrick, H. (1994) Introduction. In: Aoki, M. & Patrick, H. (eds.) *The Japanese Main Bank System*, New York: Oxford University Press.

Aoki, M., Patrick, H. & Sheard, P. (1994) The Japanese Main Bank System: An Introductory Overview. In: Aoki, M. & Patrick, H. (eds.) *The Japanese Main Bank System*, New York: Oxford University Press.

Baxter, P. & Jack, S. (2008) Qualitative Case Study Methodology: Study Design and Implementation for Novice Researchers, *The Qualitative Report* 13 (4): 544-559.

Beechler, S.L. (2005) The Long Road to Globalization: in Search of a New Balance between Continuity and Change in Japanese MNCs. In: Pudelko, M. & Haak, R. (eds.) *Japanese Management: The Search for a New Balance between Continuity and Change*, New York: Palgrave Macmillan.

Børsen. (2013) *Vestas and Mitsubishi join forces in offshore turbines* (in Danish; *Vestas og Mitsubushi går sammen on havmøller*): http://tinyurl.com/jkaamca (accessed 10 January 2017).

Børsen. (2015) *Locked Mitsubishi millions flow into MHI Vestas* (in Danish; *Låste Mitsubishi millioner er rullet ind i MHI Vestas*): http://tinyurl.com/mggzqvl (accessed 10 January 2017)

Buck, T. (2011) Case selection informed by theory. In: Piekkari and Welch (eds.) *Rethinking the case study in International Business and Management Research*, Cheltenham: Edward Elgar.

Dawani Group Holding. (2015): http://www.dawanigroupholding.com (accessed 10 January 2017).

Dong Energy. (2015) *DONG Energy signs agreement with MHI Vestas Offshore Wind for the Borkum Riffgrund 2 Offshore Wind Farm*: http://tinyurl.com/hfd6wso (accessed 10 January 2017).

Dore, R. (1994) Equality-Efficiency Trade-Offs. In: Aoki, M. & Dore, R. (eds.) *The Japanese Firm: The Sources of Competitive Strength*, New York: Oxford University Press.

Dore, R. (2005) Innovation for Whom? In: Pudelko, M. & Haak, R. (eds.) *Japanese Management: The Search for a New Balance between Continuity and Change*, New York: Palgrave Macmillan.

Dunning, J.H. (1997) *Alliance Capitalism and Global Business*, London; New York: Routledge.

The Economist. (2015) *All in the family:* http://www.economist.com/node/2714699 (accessed 10 January 2017).

Energy Watch. (2013) *Japanese government behind Vestas-MHI agreement* (in Danish; Japansk regering bag Vestas-MHI aftale): http://tinyurl.com/hjpby8m (accessed 10 January 2017).

Fletcher, M. and Plakoyiannaki, E. (2011) Case selection in international business: key issues and common misconceptions. In: Piekkari & Welch (eds.) *Rethinking the case study in International Business and Management Research*, Cheltenham: Edward Elgar.

Gerlach, M.L. (1992) *Alliance Capitalism: The Social Organization of Japanese Business*, Oxford: University of California Press.

Hobson, C. (2014) What is the role for nuclear power in Japan? *The Asia-Pacific Journal* 11(27).

Hoshi, T. (1994) The Economic Role of Corporate Grouping and the Main Bank System. In: Aoki, M. & Dore, R. (eds.) *The Japanese Firm: The Sources of Competitive Strength*, New York: Oxford University Press.

Hoshi, Kashyap & Scharfstein, (1991) Corporate Structure, Liquidity, and Investment: Evidence from Japanese Industrial Groups, *The Quarterly Journal of Economics* 106(1): 33-60.

INPEX. (2014) *Abu Dhabi's Petroleum Institute Announces Completion of INPEX-sponsored Solar Car*: http://www.inpex.co.jp/english/csr/topics/2014/e20141224-a.pdf (accessed 10 January 2017).

INPEX. (2015) *INPEX Commences Oil Production from Nasr Oil Field Offshore Abu Dhabi, United Arab Emirates*: http://www.inpex.co.jp/english/news/pdf/2015/e20150205.pdf (accessed 10 January 2017).

Japan-Saudi Arabia Industrial Cooperation Task Force. (2012) *Japan-Saudi Arabia Industrial Cooperation: Past activities, achievements and future policy*: http://tinyurl.com/hgskghl (accessed 10 January 2017).

Kantei. (2013) *Japan's Ties with the Middle East in a New Age of Synergy, Mutual Prosperity, and Cooperation*, Policy Speech by Prime Minister Shinzo Abe: http://www.mofa.go.jp/files/000004096.pdf (accessed 10 January 2017).

Kingston, J. (2014) Japan's nuclear village: Power and resilience in Kingston, J. *Critical Issues in Contemporary Japan*, New York: Routledge.

Kono, T. & Clegg, S. (2001) *Trends in Japanese Management: continuing strengths, current problems and changing priorities*, Basingstoke, Hampshire: Palgrave.

Lincoln, J. R. & Gerlach, M. L. (2004) *Japan's Network Economy: Structure, Persistence and Change*, Cambridge: Cambridge University Press.

METI. (2014) *The 2014 Strategic Energy Plan* (provisional English translation): http://tinyurl.com/jtqswje (accessed 10 January 2017).

MHI. (2010) *Annual report for the year ended March 31, 2010.*

MHI Report. (2014): http://tinyurl.com/zxlo36v (accessed 10 January 2017).

MHI. (2015): http://www.mhi-global.com/ (accessed 10 January 2017).

MHI-Vestas. (2014) *MHI Vestas Offshore Wind: Velling Mærsk Project Secures Phase Two of Development for the V164-8.0 MW Offshore Wind Turbine*: http://tinyurl.com/ztufvzq (accessed 10 January 2017).

MHI-Vestas. (2015): http://www.mhivestasoffshore.com/About/Profile/The-Company (accessed 10 January 2017).

Mills, A.J. And Mills, J.H. (2011) Digging Archaeology: postpositivist theory and archival research in case study developments. In: Piekkari, R. & Welch, C. *Rethinking the Case Study in International Business and Management* Research, Cheltenham: Edward Elgar.

Mir, R. (2011) Blurring the boundaries between case analysis and ethnography: reflections on a hybrid approach to researching multinational corporations. In: Piekkari & Welch (eds.) *Rethinking the case study in International Business and Management Research*, Cheltenham: Edward Elgar.

Mumtalakat. (2015): http://www.bmhc.bh (accessed 10 January 2017).

NAIIC. (2012) *The Diet of Japan: The Fukushima Nuclear Accident Independent Investigative Committee: The official report* (English Summary): http://www.cryptome.org/2012/07/daiichi-naiic.pdf (accessed 10 January 2017).

NRA. (2015): http://www.nsr.go.jp/english/e_nra/ (accessed 10 January 2017).

Oh, I. (2004) *Japanese Management: past, present and future*, Singapore; Tokyo: Pearson, Prentice Hall.

PI. (2015): http://www.pi.ac.ae/aboutPI.php (accessed 10 January 2017).

Piekkari, R. & Welch, C. (2011) Pluralism in international business and international management research: making the case. In: Piekkari and Welch (eds.) *Rethinking the case study in International Business and Management Research*, Cheltenham: Eward Elgar.

Porter, M. E., Takeuchi, H. and Sakakibara, M. (2000) *Can Japan Compete?* Cambridge, Massachusetts: Perseus.

Pudelko, M. (2005) Continuity vs. Change? The Key Dilemma for Japanese Management. In: Pudelko, M. & Haak, R. (eds.) *Japanese Management: The Search for a New Balance between Continuity and Change*, New York: Palgrave Macmillan.

Pudelko, M. & Haak, R. (2005) The Current State for the Japanese Economy and Challenges for Japanese Management: An Overview. In: Pudelko, M. & Haak, R. (eds.) *Japanese Management: The Search for a New Balance between Continuity and Change*, New York: Palgrave McMillan.

Reuters. (2013) *Vestas and Mitsubishi Heavy form joint venture:* http://tinyurl.com/hfo38aq (accessed 10 January 2017).

SBI Holdings. (2013a) *Memorandum of Understanding between Mumtalakat, the Sovereign Wealth Fund of Bahrain, and SBI Pharmaceuticals Co., Ltd. to Cooperate in the 5-ALA Business in Bahrain Focusing on the Pharmaceuticals Industry*: http://tinyurl.com/juya8q9 (accessed 10 January 2017).

SBI Holdings. (2013b) *Notice on the Establishment of Joint Venture Company with Dawani Group Holding, a Major Corporate Group in Bahrain*: http://tinyurl.com/jjkbvjg (accessed 10 January 2017).

SBI Pharmaceuticals. (2015): http://www.sbipharma.co.jp/english/index.html (accessed 10 January 2017).

Sheard, P. (1994) *Interlocking shareholding and corporate governance in Japan* in Aoki, M. & Dore, R. (eds.) *The Japanese Firm: The Sources of Competitive Strength*, New York: Oxford University Press.

SWIFI. (2013) *SBI Pharmaceuticals signs MoU with Mumtalakat Holdings*: http://tinyurl.com/dxa6q97 (accessed 10 January 2017).

Teramoto, Y. & Benton, C. (2005) Organizational Learning Mechanisms for Corporate Revitalization. In: Pudelko, M. and Haak, R. (eds.) *Japanese Management: The Search for a New Balance between Continuity and Change*, New York: Palgrave Macmillan.

Toyota. (2015) *Saudi Japanese Automobile High Institute*: http://tinyurl.com/j7qao98 (accessed 10 January 2017).

Uchida, H. (2013) *Japan's Energy Policy after the 3.11 Natural and Nuclear Disasters - From a Viewpoint of the R&D of Renewable Energy and its Current State*, presented as Keynote Lecture at the 3rd International Conference on Energy Process Engineering (ICEPE) 2013, "Energy Transition to Renewable Energy Systems", June 3 to 6, 2013, Frankfurt, Germany.

Vestas. (2013) *Vestas and Mitsubishi Heavy Industries form a joint venture dedicated to offshore wind energy*: http://tinyurl.com/guo3tq7 (accessed 10 January 2017).

Vestas. (2015): http://www.vestas.com (accessed 10 January 2017).

Vestas. (2014) *Press release: Final establishment of MHI Vestas Offshore Wind (in Danish; Endelig etablering af MHI-Vestas Offshore Wind)*: http://tinyurl.com/zyafny3 (accessed 10 January 2017).

Whittaker, D.H. (1994) 'SMEs, Entry Barriers and Strategic Alliance's in Aoki, M. & Dore, R. (eds.) *The Japanese Firm: The Sources of Competitive Strength*, New York: Oxford University Press.

Westney, E. (2006) The Lost Decade and and Japanese Business Studies, *Asian Business and Management* 5(2): 167-185.

Witt, M. (2006) *Changing Japanese Capitalism*, New York: Cambridge University Press.

Yin, R. K. (2009) *Case study research: Design and methods* (4th ed.). Thousand Oaks, California: Sage.

8

Proposal for the Establishment of the "Bahrain Industrial Partnership Innovation Centre"

Ebrahim Radhi

Introduction

The following paper evaluates the role of Technology Innovation Centres (TICs) in helping Bahrain evolve into a knowledge-based economy. The TIC is "*a critical element to deliver governmental or wider public sector, policies, and strategies to promote innovation*" (Hauser, 2010: 9). Generally, TICs promote various roles in addressing the challenges of technological commercialisation and in overcoming barriers of innovation. TICs are strong platforms for transnational connections and are used as knowledge transfer hubs. Bahrain's economy clearly recognizes the necessity to shift from an oil-based economy to a knowledge-based economy – especially with the ongoing fall in oil prices worldwide.

This chapter examines an emerging economy focusing on its capability to utilise TICs as knowledge-based facilitators for economic growth. A comparative assessment of Bahrain and Qatar reveals that in comparison to Qatar, Bahrain has lacked the motivation to deploy TICs for innovation and technology development. Qatar, on the other hand, has expanded TICs in many forms with substantial international connections and collaborations.

One of Bahrain's experiences toward setting up TICs is the engagement between JAFCON Consultants and the Ministry of Industry and Commerce on a unique national project called the "Subcontracting & Industrial Partnership", whose objective was to help Bahraini small and medium enterprises (SMEs) manufacture industrial products that used to be imported. The project identified 399 imported products for these Bahraini-based SMEs to manufacture, collectively worth $1.6 billion. Based on this experience, the following chapter proposes the establishment of the Bahrain Industrial Partnership Innovation Centre (BIPIC), a technology innovation centre to help Bahraini SMEs and entrepreneurs manufacture imported products in Bahrain. BIPIC intends to provide technical and other supports for localizing these imported products. BIPIC plans to be a platform for enabling transnational intellectual linkages between experts in Bahrain and elsewhere and to attract foreign direct investments.

Bahrain's vision for 2030 is an economic roadmap to develop Bahrain with the ultimate goal of enhancing the standard of living for Bahrainis. The vision is based on creating a knowledge-based economy instead of depending on an oil-based economy. This paper focuses on the core areas of a knowledge economy: technology and innovation. As acknowledged by the World Economic Forum (WEF), "*in the long run, standards of living can be enhanced only by technological innovation*" (Schwab, 2010-2011: 8).

Fagerberg (2006) states that while invention is the idea for a new product or process, innovation is the commercialisation of that idea. Inventions can be carried out in universities or laboratories by an inventor. By contrast, innovation is performed by an entrepreneur in a commercial context.

The establishment of Technology Innovation Centres (TICs) is a strategy for promoting technological commercialisation and overcoming barriers of innovation. In fact, early government estimates suggested that Bahraini TICs would significantly contribute to the objective of increasing industrial output in Bahrain's GDP from 12% in 2008 to 36% by 2018 (Al-A'ali, 2009).

The manufacturing sector is considered to be one of the most promising sectors in Bahrain. In 1992, the United Nations Industrial Development Organization (UNIDO) mission report suggested that the future backbone of Bahrain's economy depended on manufacturing value added activities (UNIDO, 1992). This is attributed to the strategic location of Bahrain, its upstream and downstream manufacturing businesses, and its business friendly environment. Moreover, the manufacturing sector is of strategic importance to diversify Bahrain's economy and make it less dependent on hydrocarbons. The Economic Development Board selected the manufacturing sector, along with finance, higher education, tourism, logistics, and information and communication technology (ICT), as focal points for future economic growth in Bahrain.

As an indication of the importance of manufacturing in Bahrain, Bahrain is now investing in more than $10 billion in new industrial zones, ports, airports, transportation infrastructures, and steel. The non-oil sector includes petrochemicals, ship repair, aluminium refining, and light manufacturing (Passport Report, 2014).

The primary aim of this research is to understand the role of TICs in supporting the goals of Bahrain's vision for 2030. To achieve this objective, this chapter will focus on the establishment of TICs as an innovation policy. To investigate this topic three research questions have been identified: What are the characteristics of TICs? How would TICs fit within Bahrain's innovation strategy? What lessons can Bahrain learn from the development of TICs in Qatar? Data for this research project was drawn from a variety of sources including case studies, interviews, and government reports.

Technology and Innovation in Bahrain

Although the economic diversification process started in the early seventies, Bahrain's economy is centred primarily on its oil and aluminium industries. One of Bahrain's early industrial achievements is its aluminium production; Bahrain is the only GCC country that uses 60% of its aluminium smelting output domestically in middle and downstream manufacturing.

However, Bahrain's economy is relatively diverse as evidenced by how the following sectors contribute to its GDP: petroleum (20.4%), finance (16.6%), manufacturing (14.4%), government services (12.8%), and other sectors that contribute less than 10% including: real estate, construction, trade, information and communication technology, and social services (CIO, 2014). Notwithstanding the growing complexity of the Bahraini economy, the emergence of these non-petroleum based sectors has failed to protect the economy from large fluctuations in oil prices (Jaffari, 2004). Therefore, transitioning to a knowledge-based economy continues to be an urgent priority for the Bahraini government.

In 2010, the Bahrain Economic Development Board organized a conference called "Innovation Bahrain" with the objective of formulating new innovative policies. The aim of these policies was to encourage the use of innovation and technology on a wider scale due to its projected impact on the local economy and on Bahrain's ability to be competitive.

In 2011, although the government had made significant strides in technological acquisition scoring (22nd) in worldwide rankings according to the Global Competitive Report, Bahrain ranked poorly in aptitude for innovation (67th), quality of scientific research institutions (117th), company spending on R&D (101st), and university-industry collaboration in R&D (88th). These factors by themselves represent the reasons for low innovation in Bahrain. Comparing Bahrain's rankings in 2011 and 2014 revealed that it is not improving in terms of innovation, but rather its capacity to innovate had deteriorated (The Global Competitiveness Report, 2011+2014).

To assess the weaknesses in Bahrain's innovation strategies, the current policies were compared to the generic innovation policies framework developed by Garofoli and Musyck (2003). The analysis revealed that one of the inadequacies was the minimal use of Technology Innovation Centres. Aside from the lack of TICs, the analysis also demonstrated that the innovation deficit in Bahrain is exacerbated the by the following deficiencies: the absence of intellectual property laws, little collaboration between universities and industry, a weak culture of entrepreneurship, the absence of technology and R & D centres, and inadequate funding for R&D on the part of large companies (EDB, Bahrain Economic Quarterly Q2-Q3 2010, 2010).

Technology Innovation Centre (TIC)

There is no consistently accepted definition of a TIC. Several terms are used to describe similar establishments, such as Research Park, Technology Park, Business Park, and Science Park (Dettwilera et al., 2006).

The TIC is a centre that promotes various roles in addressing the challenges of technological commercialisation and in overcoming barriers of innovation. The definition of TIC is "*a critical element to deliver governmental or wider public sector, policies and strategies to promote innovation*" (Hauser, 2010: 9). TIC priorities vary widely and are largely context-dependent with some roles receiving more importance than others (Hauser, 2010). Typical roles include: offering technical consultancy services, facilitating technology transfers, the commercialisation of innovative ideas and technologies, the creation of Spin-off companies, incubation and R&D support, attracting foreign direct investment, and linking SMEs with large firms to act as their suppliers.

TICs also have divergent models of funding, operation, and assessment. The funding model varies significantly among countries; however, there are three broad sources of funds: government funds, research funds from organizations, and private sector funds (Hauser, 2010). The operational models are highly dependent on the sector and/or technology that the TIC supports. Human resource development, which aims at increasing technical and applied engineering skills, is a key role that TICs undertake either explicitly or implicitly. TICs are typically assessed by their economic outcomes. As such, effectiveness is determined by their ability to generate economic activities that outweigh investment, to stimulate a vibrant industry, and to maintain industrial partners and customers who will invest time and money (House of Commons, 2011).

In addition to these models, key consideration must be given to stakeholders. TICs create collaborative synergies with universities and private and public sectors to achieve their intended goals. The stakeholders vary depending on the type of the TIC. TICs that are concerned with novel technologies usually partner with universities. TICs that focus on technologies demanded by the market mostly partner with business firms.

It is vital that TICs have a clear vision that aligns with the national development strategies of the country in which they operate (Hauser, 2010). TICs should have a high level of autonomy and independence in order to adapt and fit the targeted sector or technology (Hauser, 2010; House of Commons, 2011; TSB, 2011). Another necessary element is to establish innovative initiatives that encourage public sector and governmental enterprises to run procurements in the technology areas where TICs are active, thus helping them to commercialise their ideas and research (Lundvall and Borras, 2005; Hauser, 2010).

TICs generally focus on a particular sector or technology (House of Commons, 2011). The sector-focused-TICs establish a group of complementary disciplines and/or are a part of a value chain within the sector (Hauser, 2010). In the selection process of sectors or technologies, it is required to consider national endowments and capitalise on existing local resources. However, such resources should not limit the possibility of TICs implementing other potential emerging sectors (Mina et al., 2009). For example, the fact that manufacturing currently represents 14% of Bahrain's GDP should not preclude Bahraini efforts to establish TICs for the manufacturing sector.

Qatar Case Study

Qatar was selected for this case study because it is capitalising on TICs in order to achieve its national vision for 2030. Additionally, Qatar possesses a common culture with Bahrain in that they have followed similar economic development paths. However, the major difference between the two countries is the wealth gap; Qatar's GDP is $83.9 billion with a GDP growth of 16.3%, while Bahrain's GDP is only $20.2 billion with a growth rate of 4.4% in 2011. Furthermore, Qatar is ranked by the WEF (2010-2011) as the second highest MENA country in terms of innovation and the 23rd worldwide out of 133 countries. And, importantly, it is ranked as the top country for procuring advanced technology products. The quality of its scientific research is very high (22nd), and it is now ranked 9th worldwide based on the number of scientific and engineering think tanks.

As shown above, although Qatar's GDP and GDP growth is four times greater than Bahrain's, global competitiveness rankings indicate that aside from innovation and sophistication, the two are relatively similar. Qatar is ranked significantly higher than Bahrain in most of the indicators for innovation. For example, Qatar is ranked 22nd in terms of the quality of scientific research institutions whereas Bahrain is ranked 117th (Schwab, 2010; 2011).

The above ranking indicates that Qatar is considerably ahead of Bahrain in company spending on R&D, university-industry collaboration in R&D, and the quality of scientific research institutes. This result reflects the fact that companies in Bahrain, even large national companies, tend not to invest in R&D. Also, it indicates that universities in Bahrain play an inadequate role in producing scientific research as well as in collaborating with industry.

Much like Bahrain, Qatar has formulated a national vision for 2030. This vision was authored by the General Secretariat for Development Planning, which is equivalent to the Economic Development Board in Bahrain. Qatar's national vision has four main objectives: human, economic, social, and environmental development.

Human and economic development focuses on the importance of innovation and education as well as establishing an effective system for scientific research. The establishment of TICs has been identified explicitly as a method of achieving both human and economic development, especially in cooperation with specialized and international leading research centres (GSDP, 2011).

Qatar Foundation (QF) was established in 2005 as the primary organisation for achieving human and economic development in Qatar. QF's mission is "*to make Qatar a leader in innovative education and research. To achieve that mission, QF supports a network of centres and partnerships with elite institutions*" (QF, 2011). To make this mission a reality, QF has implemented numerous initiatives on its site, such as establishing the Strategic Research Board that oversees the development of science, education, and technology in Qatar.

QF has attracted nine international universities in an effort to build Qatar Education City. Moreover, QF has sponsored the establishment of four research institutes. Both the universities and the institutes encourage international companies to establish TICs in Qatar Science and Technology Park (QSTP) such as: $50m GE R&D-Centre, ExxonMobil Technology Centres, and Rolls-Royce Centre (QSTP, 2011).

QSTP offers attractive premises and facilities, a free zone, and consultation services for advice on investing in Qatar. Moreover, QSTP focuses on the commercialisation of new technologies by supporting TICs through direct funds and various supportive programs (QSTP, 2011).

QF creates a collaborative and intellectually interactive environment among universities, research institutes, and companies. To extract the maximum benefit from this rich collaboration, QF formed the Strategic Research Board. One of the board's principal agents is the RAND-Qatar Policy Institute, which helps make improvements in Qatar's policy and decision making in education, science, and community development (RQPI, 2011). Another branch is QF's research and science programs. These programs work as a network with the universities, research institutes, companies, and R&D centres.

The third branch is the Qatar National Research Fund (QNRF), which fosters a research culture that "*aims to advance knowledge and education by providing support to researchers*" (QNRF, 2011). QNRF tries to achieve its aim through (1) encouraging collaborative research among academia, research centres, and private and public sectors inside and outside of Qatar, (2) funding research at universities, and (3) enhancing the education of Qatar's citizens and the training of its workforce.

In accordance with the House of Commons TIC classification typology (2011), four Qatari TICs were selected for review: Qatar Robotic Surgery Centre, a public sector research establishment TIC, Qatar Mobility Innovations Centre, a university spin-out TIC, Qatar Carbonates and Carbon Storage Research Centre, an independent research and technology organization, and Chevron's Centre for Sustainable Energy Efficiency and GreenGulf, both of which represent the corporate research TIC model.

Qatar Robotic Surgery Centre (QRSC) is a collaborative initiative between Hamad Medical Corporation – Qatar's largest public hospital – and Qatar Science and Technology Park. "QRSC has three main purposes: demonstrate new medical robots, train surgeons from Qatar and the region, and participate in developing robotic surgery technology". QRSC is engaged in collaborative research with robot manufacturers and medical surgery centres worldwide. It aims to transform Qatar into a regional hub of excellence in training, development, and clinical practice in robotics surgery (QSTP, 2011).

Qatar Mobility Innovations Centre (QMIC) is an extension from Qatar University and was established within Qatar Science and Technology Park in 2009. QMIC aims to be "a platform for carrying out applied research, providing consulting services, promoting development of human assets, and championing educational and training activities in the area of wireless systems and applications" (QUWC, 2011). QMIC plans to achieve its aims through establishing a strong industry-government-university partnership, institutional partnership agreements with world leading companies, institutions and research centres, and applied research in a number of key industries in Qatar including oil and gas, healthcare, environment, transportation, security, utilities, media, and sports (QMIC, 2011).

Qatar Carbonates and Carbon Storage Research Centre (QCCSRC) represents an international collaboration, and it is positioned outside of Qatar. As its name suggests, it "investigates challenges in the exploitation of carbonate reservoirs by combining the international expertise of Shell, the local knowledge of Qatar Petroleum, the assistance and strategic advice of Qatar Science and Technology Park (QSTP), and the research strengths of Imperial College London, through its Energy Futures Lab" (QCCSRC, 2011). It is a $70M, 10-year research centre jointly funded by Qatar Petroleum, Shell, and QSTP. Its research includes "reservoir simulation and experimental studies at laboratory, pilot and field scales" (QCCSRC, 2011). QCCSRC aims to develop new approaches for extracting oil and gas (Qatar endowments) by integrating and optimising evolving relevant technologies.

Moreover, QCCSRC practices capacity building and knowledge transfer to Qatar through training from the Imperial College of London for PhD students and postdoctoral researchers, transferring techniques and methodologies from the Imperial College of London to Qatar's industry, and linking them with local universities in Qatar (Macdonald, 2011).

Even though QCCSRC is located outside of Qatar, it is connected with Qatari Universities and businesses through academic activities and joint research. It also provides financial support in terms of research grants (Macdonald, 2011). QCCSRC focuses on a specific sector while having a high level of autonomy in order to adapt to fit the sector. It also believes in having an interactive collaboration among its stakeholders and in the continuous development of its employees (Macdonald, 2011).

Corporate research functions and laboratories such as Chevron's Centre for Sustainable Energy Efficiency (CSEE) and GreenGulf CSEE focus on three promising technologies for Qatar: solar power, solar air conditioning, and energy efficient lighting that work best in Middle Eastern conditions (Chevron, 2009). CSEE is the first Chevron Energy Solutions (CES) project outside of the U.S. It aims to research the development and application of renewable energy through collaboration with universities in Qatar. CSEE is one of the main initiatives Qatar is taking to become a major clean technology contributor.

In April 2010, CESS and GreenGulf (a Qatar-based renewable energy company) created Chevron Qatar Energy Technology, investing $20 million to perform joint studies to test solar energy technologies and their applications in the Middle East (GreenGulf, 2010).

GreenGulf benefits from being at QSTP because it leads to easier access to government projects and to partnerships with international reputed companies such as Chevron. GreenGulf utilises the partnership with Chevron though conducting joint projects as well as sending GreenGulf's employees to training and development programs delivered by Chevron. GreenGulf finds having a clear vision aligned with Qatar's national vision is of paramount importance. It also believes in creating a collaborative environment and establishing international bonds (Al-Kuwari, 2011).

Discussion and Recommendations

The Bahraini government's manufacturing strategy selected TICs as one of the instruments that would contribute significantly to boost the manufacturing sector in Bahrain (Al-A'ali, 2009). Therefore, in order for Bahrain to benefit from the advantages of TICs, Bahrain has to confront many challenges in the field of innovation, as reflected by its low WEF ranking in 2011 and 2014.

The main reasons that Bahrain ranked low in innovation have to do with a lack of technology and R&D centres, the absence of funding bodies, the scarcity of qualified researchers, the poor role that the universities play, inadequate intellectual property laws, and limited R&D initiatives in companies (EDB, 2010). The above reasons are considered the main hindrance for TICs success. Accordingly, for TICs to fit Bahrain's innovation strategy, Bahrain is required to take tangible steps towards engendering innovation in its economy. Therefore, it is prudent to assume that the removal of the negative contributors to Bahrain's low innovation international ranking will, by default, provide greater impetus for TICs to be successful.

Bahrain can draw important lessons from Qatar's track record of developing TICs. Qatar's national vision clearly stressed the viability of TICs in the march towards technological advancement. The development of TICs is embedded in the comprehensive national science, education, and community development strategies lead by Qatar's Strategic Research Board and developed by RAND-Qatar Policy Institutes, which is considered a specialised policy development institute. According to the case studies analysis, there are strong bonds between Qatar's national vision and the missions of the TICs in Qatar. The majority of these TICs target Qatar's market needs by either building on Qatar's resources (Oil & Gas) or building the infrastructure of knowledge-based economy in Qatar.

With the goal of enhancing the science and research development, Qatar invites international universities to conduct research and form strong university – industry bonds, both inside and outside of Qatar. It also attracts international companies to place their R&D facilities in Qatar by offering attractive facilities and encourages local companies to engage in developing new technologies and setting up TICs. The wealth of Qatar was arranged for this purpose. For Bahrain, a gradual approach is needed as the capital for such investments are not available.

Furthermore, QF's campus was designed to host the universities at Qatar Education City and to host the TICs of companies and other research centres in Qatar Science and Technology Park. QF encourages collaborations and synergies between the entities of the city and the park through collaborative research and science programs. These programs operate as a network involving the universities, research institutes, and companies' TICs.

To achieve this objective QF provides funds through Qatar National Research Fund. Qatar National Research Fund encourages the culture of research through funding programs, embracing schools, universities, private and public sectors, and collaborative joint research. Moreover, QSTP offers premises, facilities, supportive services, and programs.

Equally important, Qatar focuses on the development of local Qataris in a quest to produce national scientists and researchers through financial support, advanced training, and education both inside and outside Qatar.

Based on the findings of this research, to develop TICs in Bahrain, it is recommended that Bahraini authorities consider the following proposals:

1. Bahrain's innovation strategy should incorporate the expansion of TICs. The mission of TICs should be linked to Bahrain's vision for 2030. These TICs should serve selected technologies and sectors that maximise Bahrain's resources (oil and gas) or focus on promising emerging technologies and manufacturing sectors.
2. Bahrain should encourage developing different types of TICs with an emphasis towards "market pull" types. This is because of the strong presence of industry and the weak linkages with the universities in Bahrain. The private sector and universities should be involved in the development of these TICs and be represented in management boards.
3. Bahrain should take advantage of TIC supporters of success including the establishment of research funding and the enforcement of strong intellectual property laws.
4. TICs should be supported by establishing innovative public procurements, such as buying the technologies developed by these TICs to foster their commercialisation.
5. TICs in Bahrain should support SMEs to enable them to provide product and services for large firms that could replace the larger firms' international purchases.
6. Companies in Bahrain should be encouraged to open R&D centres, especially large enterprises, and to work with SMEs and universities to stimulate innovation.
7. TICs in Bahrain should form international alliances and partnerships to gain access to basic research and unique technologies that could be commercialised through TIC's. These alliances should be with international universities, multi-international companies, and leading TICs.
8. Bahrain should provide facilities and funds to stimulate collaborative research between companies, universities, and research institutes inside and outside Bahrain. This can be achieved through collaborative research programs such as in Qatar.

Subcontracting & Industrial Partnership (SIP)© Pilot Project

One step toward implementing recommendation number five of the above section was implemented through the Ministry of Industry & Commerce; in 2011, the ministry initiated the Subcontracting & Industrial Partnership (SIP) Pilot Project in cooperation with JAFCON Consultants. The prime objective of this project was to enhance the national capacity of the manufacturing sector in Bahrain. SIP specifically aims at exploring the potential of local small & medium enterprises (SMEs) to produce industrial and engineering components currently imported from overseas to large domestic firms (LDFs) without compromising quality – the products had to meet international standards and specifications (Jafcon, 2014).

The essence of the JAFCON and SIP's framework is to make a successful match as its final result. This implies that the product, which is currently imported from outside, can be locally manufactured at the same or at higher specifications.

The first phase was to identify imported industrial components, which was achieved through deploying the Purchase Information System (PIS). The industrial components then were evaluated through the Value Added Screening (VAS) and Import or Enable Making (I/EM) to ensure that these components were of value added and could feasibly be manufactured in Bahrain.

The readiness and ability of SMEs' resources were examined through the Resources Gap Analysis (RGA). The aim of RGA is to equip the selected SMEs with the necessary technical, technological, and financial support to provide selected industrial products and services at the required quality standards, i.e. becoming qualified SMEs. Productivity Health Checks (PHC) were then applied over the SMEs' to ensure their sustainability over the long term.

The process of matching then started, which enabled SMEs to manufacture the selected industrial components for LDFs. After communicating with LDFs to gain prototype approval, SMEs eventually became subcontractors for LDFs.

The Resource Gap Analysis during the study covered 9 manufacturing segments including 32 SMEs, which represents the manufacturing sector in Bahrain. The main findings were as follows:

- There are numerous components that LDFs require for operations, but lack of technical know-how and a lack of advanced technology were the main obstacles for SMEs to manufacture these components locally.
- The lack of communication between LDFs and SMEs was found to be a major obstacle.
- The committed SMEs were ready to invest and grow their capacity to meet the LDFs requirements, but they needed long term contracts with adequate incentives, which are not provided by LDFs.
- There were almost no financial gaps for SMEs.

The project successfully determined a $1.6 billion volume of business for major LDFs in Bahrain. The target of the pilot study was to manufacture 12 components; however, due to the availability of high opportunities, the number of manufactured components increased from 12 to 399, spreading over 32 categories of imported components.

More than 30 manufacturing SMEs were enabled to manufacture these components at the same or higher standard than the required specifications. This was a reflection of the larger goal of enhancing their overall performance and capacity.

Moreover, the locally manufactured components secured significant savings compared to the cost of importing. The cost savings in some components were up to 98.1%, with an overall savings of 60% for the 399 products studies. Adding to the savings, LDF benefits include higher operational security and an increased flexibility in terms of the speed of delivery and the flexibility of ordering.

Last but not least, SIP successfully attracted foreign investments into Bahrain, where a factory was specially established to locally manufacture the imported products for LDFs.

In order to sustain the success of the SIP project, this chapter proposes the establishment of a Bahrain Industrial Partnership Innovation Centre (BIPIC).

The Proposed BIPIC is a physical Technology Innovation Center, where imported industrial products to Bahrain are identified with the aim of enabling Bahraini SMEs/ entrepreneurs to manufacture these imported products locally. BIPIC intends to provide technical and other support to localize these imported products – as the main gaps were found to be technical in nature.

BIPIC shall serve not only the major large domestic firms but will also cover governmental organizations and international companies in Bahrain.

BIPIC's anticipated outcomes are as follows:

- Become a platform for knowledge transfer within Bahrain as well as with international companies and knowledge transfer agencies.
- Provide entrepreneurs with value-added business ideas and enable them to establish their manufacturing start-ups.
- Assist SMEs to expand by locally manufacturing imported products.
- Attract Foreign Direct Investment (FDI) to Bahrain
- Establish joint ventures between FDI and local Entrepreneurs/SMEs which will lead to knowledge transfer and create jobs.
- Reduce reliance on imported products and, in so doing, limit capital expatriation.

Conclusion

Technology and Innovation Centres address the challenges of technology commercialisation and help in overcoming barriers to innovation. TICs are strong platforms for transnational connections and are used as knowledge transfer hubs.

Bahrain is in need of transforming its economy away from an oil-based economy towards a knowledge-based economy. One internationally proven tool to foster innovation and entrepreneurship is the TIC. Therefore, Bahrain's case was examined for its capability of utilising TICs as a means of enhancing local innovation. Qatar has made a remarkable leap in utilizing TICs. Given the similarities between the two nations, Bahrain is likely to succeed if it follows a similar model.

In Bahrain, JAFCON consultants worked with the Ministry of Industry and Commerce on a unique national project to support SMEs to locally manufacture industrial products that had previously been imported.

On the basis of this successful plan, my chapter proposes the establishment of the Bahrain Industrial Partnership Innovation Centre (BIPIC). BIPIC is designed to be a physical Technology Innovation Centre, where Bahrain's imported industrial products are identified and brought to the attention of Bahraini SMEs/entrepreneurs so that they can be manufactured locally. BIPIC will serve as a platform for transnational, knowledge-based relations connecting private and public sector actors in Bahrain with counterparts in other parts of the world.

References

A*STAR. (2011). *Agency for Science, Technology and Research (A*STAR)*. Retrieved 4 21, 2011, from: http://www.a-star.edu.sg/tabid/863/default.aspx

Al'aali, H. (2009, 11 1) Manufacturing Strategy . *Aswaq News* , p. 2.

Al-Khalifa. (2010, 4 7). *H.E. Sh. Mohammed bin Essa Al-Khalifa Speech*. Retrieved 4 12, 2011, from Bahrain Econmic Development Board: http://tinyurl.com/jhhgnbx

Al-Kuwari, O. (2011, 6 6) Managing Director of GreenGulf. (Ebrahim, Interviewer).

Arnold, E., Barker, K., & Slipersaeter, S. (2009). *Analysis of Public Research Institutes in Europe in Selected S&T fields: historical evolution and future scenarios literature review (revised)*. Brighton: Technopolis, Mlolr, NIFUSTEP.

Asbl, E. N. (2008). *Activities of EU member states with regard to the reform of the public research base*. Brighton: Technopolis Group.

Asheim, B. T., & Gertler, M. S. (2005). The Geography of Innovation: Regional Innovation Systems. In J. Fagerberg, D. C. Mowery, & R. R. Nelson (Eds.), *The Oxford Handbook of Innovation*. Oxford: Oxford University Press.

Åström, T., Eriksson, M.-L., Niklasson, L., & Arnold, E. (2008). *International Comparison of Five Institute Systems*. Bredgade: Forsknings- og Innovationsstyrelsen.

Bahrain Development Board (2011). Retrieved 4 17, 2011, from Bahrain Development Bank: http://www.bdb-bh.com/

Brem, A., & Voigt, K.-I. (2009). Integration of market pull and technology push in the corporate front end and innovation management – Insights from the German software industry. *Technovation* , *29*, 351–367.

Chevron. (2009). *Press Releases*. Retrieved 05 26, 2011, from Chevron: http://www.chevron.com/news/press/release/?id=2009-02-17

Al-Khalifa. (2010, 4 7). *H.E. Sh. Mohammed bin Essa Al-Khalifa Speech*. Retrieved 4 12, 2011, from Bahrain Econmic Development Board: http://tinyurl.com/jhhgnbx

Central Inforamtion Organization (CIO). (2014) – *Bahrain Open Data*. Retrieved 1 12, 2014 from Central Inforamtion Organization: http://www.data.gov.bh/

Dee, N., Ford, S., & Garnsy, E. (2007). Development and Commercialisation of Eco-Innovations by New Ventures. Cambridge: Center for Technology Management, Univeristy of Cambridge.

Dettwilera, P., Lindelofb, P., & Lofstenc, H. (2006). Utility of location: A comparative survey between small new technology-based firms located on and off Science Parks – Implications for facilities management. *technovation* , *26*, 506–517.

Dyson, J. (2010) *Making the UK the leading high tech exporter in Europe* . London : James Dyson.

Econmic Development Board (EDB). (2011). *About*. Retrieved 4 12, 2011, from Bahrain Economic Development Board: http://www.bahrainedb.com/m_AboutEDBIntro.aspx

Fagerberg, J. (2006). Innovation a guide to the literature. In *The Oxford Handbook of Innovation*. Oxford, UK: Oxford University Press.

Garofoli, G., & Musyck, B. (2003). Innovation policies for SMEs: an overviw of policy instruments. In B. Asheim, A. Isaksen, C. Nauwelaers, & F. Todtling (Eds.), *Regional Innovation Policy for Small-Meduim Enterprises*. Cheltenham: Edward Elgar.

GreenGulf. (2010). *Media centre*. Retrieved 05 16, 2011, from GreenGulf: http://tinyurl.com/zo568qg

General Secretariat for Development Planning (2011) *Qatar National Vision 2030*. from General Secretariat for Development Planning.

Hauser, H. (2010) *The Current and Future Role of Technology and Innovation Centres in the UK*. London: Hermann Hauser.

House of Commons (2011) *Technology and Innovation Centres*. Science and Technology Committee. London: The Stationery Office Limited.

Jaffari, A. (2004) *Performance of Manufacturing SMEs in Fluctuating single product oil economy - The case of Bahrain* . Seiten: VDM Verlag Dr. Müller e.K.

Jafcon Consultant for Productivity Improvement, SIP Rerport, 2014.

Lesser, S. (2011). *Top Ten Highlights of Cleantech in Qatar*. Retrieved 05 27, 2011, from CleanTechies: http://tinyurl.com/z6v39jh

Lundvall, B.-A., & Borras, S. (2005). Science, Technology and Innovation Policy. In J. Fagerberg, D. C. Mowery, & R. R. Nelson (Eds.), *The Oxford Handbook of Innovation* . Oxford: Oxford University Press.

Macdonald, I. (2011, 06 08) Program Manager, QCCSRC. (E. Radhi, Interviewer).

MET-IfM. (2010). *Understanding the industrial success of Singapore and Malaysia*. Cambridge: University of Cambridge Institute for Manufacturing (IfM).

Mina, A., Connell, D., & Hughes, A. (2009). *Models of technology development in intermediate research organisations*. Cambridge: Judge Business School, University of Cambridge.

Moir, D. (2011, 4 5) St. John's Innovation Centre. (E. Radhi, Interviewer).

Nooteboom, B. C. (1992). The purpose and effectivness of technology transfer by government-sponsored Innovation Centres. *Technology Analysis and Strategic Management* , *4* (2), 149-166.

Peters, L., Rice, M., & Sundararajan, M. (2004). The Role of Incubators in The Enterpreneurial Process. *Jornal of Technology Transfer, 29*, 83-91.

Qatar Carbonates and Carbon Storage Research Centre QCCSRC(2011) *About*. Available at: http://www3.imperial.ac.uk/qatarcarbonatesandcarbonstorage

Qatar Foundation (QF) (2011) *Universities & Postgraduate study*. Available at: http://www.qf.org.qa/

QNEXUS (2011) *R&D*. Available at: http://www.qnexus.net/rd/

Qatar National Research Fund (QNRF) (2011) *Welcome to Qatar National Research Fund*. Available at: http://www.qnrf.org/

Qatar Science and Technology Park (QSTP) (2011) *Current Memebers* from Qatar Science & Technology Park.

RAND-Qatar Policy Institute (RQPI) (2011) *About RQPI*. Available at: http://www.rand.org/qatar/about.html

Schwab, K. (2010-2011) *The Global Competitiveness Report*. Geneva: World Economic Forum.

Tamkeen. (2011). *Private Sector Support*. Retrieved 4 12, 2011, from Tamkeen: http://www.tamkeen.bh/en/

Technology Strategy Board (TSB) (2011) *Technology Innovation Centres: a prospectus*. Swindon: Technology Strategy Board.

Tunzelmann, N. V., & Acha, V. (2005). Innovation in "low tech" Industries. In The Oxford Handbook of Innovation. Oxford: Oxford University Press.

United Nations Industrial Development Organization UNIDO. (1992). *Overview of the industrial sector, industrial stratgy options and integrated programmes*. Manama: UNIDO.

9

A Preliminary Study on Strategies of Chinese Firms Collaborating with Gulf Nations' Firms

Gao Xudong

Introduction

Although China is an important oil and gas producer in the world (in 2014 domestic oil production is 210 million tons, and natural gas production is 133 billion m^3), it now relies heavily on importing oil and gas to support its fast economic development. Specifically, in 1993, China became a net importer of oil. Since then, China's reliance on importing oil and gas has been increasing rapidly (Table 1). In 2014, China's oil consumption was more than 500 million tons, almost 60% was imported; natural gas consumption was more than 183 billion m^3, about 32% was imported.

Within this context, major Chinese oil and gas firms, namely, CNPC, Sinopec, and CNOOC, are becoming increasingly active in collaborating with international partners. For example, in 2014 CNPC and its partners produced 127.47 million tons of oil equivalent, of which CNPC's share was 65.20 million tons. In 2014, CNPC also signed a purchase and sales contract with Gazprom on gas imports from Russia, under which Russia will supply China with 38 billion cubic meters of natural gas per year for 30 years from 2018. CNPC also signed a contract with Novatek to purchase 3 million tons of LNG annually for 20 years.

Table 1: Crude Oil Import in China (Million Tons)

2000	59
2005	127
2010	235
2013	282
2014	308

Source: composed by author

Collaborating with Gulf nations firms is of special importance for China (Table 2). For example, in 2014, more than 50% of imported oil is from the Persian Gulf countries, with Saudi Arabia accounting for 16.11%, Oman 9.65%, Iraq 9.27%, and Iran 8.91%. In addition, Chinese firms are also playing important roles in oil and gas exploration and development in these countries.

Given the importance of the Persian Gulf region, in this chapter, we explore the key characteristics of Chinese firms' strategies in collaborating with Gulf nations firms. Our research methodology is mainly an interview-based case study (Yin, 1989). In the past 10 years we have been working closely with Chinese oil and gas companies, especially CNPC, and have had opportunities to interview many of their top managers, engineers, and workers (we have published a book that contains 24 case studies of technological innovation at CNPC, and a new book is just finished) (Liu et al., 2007).

Table 2: Top 10 Countries Exporting Oil to China (2014)

Ranking	Countries	Million Tons of Oil
1	Saudi Arabia	49.7
2	Angola	40.6
3	Russia	33.1
4	Oman	29.7
5	Iraq	28.6
6	Iran	27.5
7	Venezuela	13.8
8	United Arab Emirates	11.7
9	Kuwait	10.6
10	Colombia	10.1

Source: composed by author

In preparing this paper, we interviewed more people in these firms, focusing on their exploration and development activities in the Gulf nations. In order to do good case studies we also draw on related streams of literature (mainly international relations in the oil and gas industry, technological innovation, firm strategy, and international businesses) (Amsden, 2001; Hennart, 2009; Hymer, 1976; Johanson and Vahlne, 2009; Price-Smith, 2015), although we are not constrained by any specific theory in our analysis.

From our study, several interesting findings emerged. First, Chinese oil and gas firms are taking on high risk ventures in collaborating with Gulf nations firms. However, they are able to lower the risks by adopting various strategies. This is easier to understand when energy partnership is considered as part of an even larger collaborative enterprisemovement. There exist strategic complementarities between China and the Gulf nations (China's fast expanding economy not only needs investment but also creates a lot of opportunities for investment while Gulf nations are transforming their economies and seeking investment opportunities globally). Collaboration between Chinese oil and gas firms and Gulf nations firms could take advantage of these complementarities.

Second, Chinese oil and gas firms are able to make important contributions in collaboration because of their increased capabilities, including the technological capabilities and capabilities to offer total solutions. Low-cost labor is becoming less and less important. For example, accordingly to our study, CNPC, the third largest oil and gas firm in the world (2014), and the 4th largest firm in the Fortune 500, now are among the largest R&D spending oil and gas firms, and have developed many leading technologies in the world.

Third, there exist big challenges in the collaboration between Chinese oil and gas firms and Gulf nations firms. In many cases, these challenges do not necessarily come from the collaborating partners but from the complicated international environments. The collaboration between Chinese firms and Iraq is a good example. In this case, the major challenge comes mainly from the Iraq-US relationship.

Theoretical Background

In this paper, we draw on several streams of literature to help with our study of the key characteristics of Chinese firms' strategies in collaborating with Gulf nations firms. First, given the special nature of the oil and gas industry, the international relations literature is highly important for this study (Bower, 2009; Stoke and Raphael, 2010; Yetiv, 2011). For example, for a long time, US and EU countries, especially the US, played a dominant role in the development of oil and gas industry in the world. Price-Smith argues that the liberal perspective fails to pay enough attention to the many conflicts among the stakeholders in the oil and gas industry, and is in need of some fundamental revision. A shadow liberalism perspective, which recognizes the importance of international political factors, is suggested to understand the evolution of the oil and gas industry in the world (Price-Smith, 2015).

Within the international relations literature, there are scholars seeing the increasing importance of not only the big powers such as the US but also other countries and organizations. Not satisfied with the influence of the US and IOCs, many countries, especially developing countries have adopted the policy of supporting their NOCs. In fact, NOCs now control most of the oil and gas reserves (Goladthau and Witte, 2010; Sampson, 1991).

In addition, international institutions such as the Organization of the Petroleum Exporting Countries (OPEC), and the International Energy Agency (IEA) also play very important roles in governing the world oil and gas industry, including influencing the prices and production levels, although these organizations also face a lot of challenges, such as coordinating member activities and managing boundaries between members and nonmembers when important stakeholders such as China and India are not included as members (Price-Smith, 2015).

From the perspective of this complicated stream of literature, it seems that Chinese firms should be able to find opportunities to collaborate with host country firms in the Persian Gulf region. First, the essence of China's foreign policy is to respect other countries' sovereignty and avoid intervening in other countries' internal affairs, especially through using military power, and promoting international collaboration. Second, China has been promoting a free trade based international economic order, especially after joining the WTO in 2001. This is important when the production and the consumption of oil and gas are asymmetrical (Table 3). Specifically, Persian Gulf countries as producers are interested in selling their oil and gas in the international market, and China as a buyer is interested in importing oil and gas from Persian Gulf countries. Both sides would be winners in the oil and gas trade. This is helpful for creating favorable conditions for Chinese firms to develop collaborative business relationships with firms in this region.

However, the Persian Gulf region is also full of uncertainties in both international relations and domestic affairs, including the US policy of deploying massive military powers in some cases. This means that doing business in this region is highly likely to take a lot of risks.

Table 3: World Oil Production and Consumption

Ranking	Proven Oil Reserves	Oil Production (2014)	Oil Consumption (2014)
1	Saudi Arabia (267)	Saudi Arabia	US
2	Venezuela (211.2)	Russia	China
3	Canada (173.6)	US	Japan
4	Iran (151.2)	Canada	India
5	Iraq (143.1)	China	Russia
6	Kuwait (104.0)	Venezuela	Saudi Arabia
7	United Arab Emirates (97.8)	United Arab Emirates	Brazil
8	Russia (60.0)	Iran	Germany
9	Libya (47.1)	Kuwait	South Korea
10	Nigeria (37.2)	Iraq	Canada

Source: composed by author

The second stream of literature is about the conditions and processes of multinational enterprises (MNEs) going global (here we put the three Chinese firms studied in this paper in the category of MNEs, although they are from China, a developing country). The first perspective in this stream of literature is mainly based on the study by Stephen Hymer (1976), whose central argument is that MNEs entering a foreign market must possess an internally transferable 'advantage' or superior resources, which could offer quasi-monopolistic opportunities to outcompete local firms in host countries (Bartlett and Ghoshal, 2002; Buckley and Casson, 1976, 1988; Chandler and Redlich, 1961; Dunford et al., 2010; Dunning, 1992, 1994; Meyer and Yen, 2006). For example, Buckley (1996) has argued that MNEs have the capability to adjust their foreign market strategies, which could increase their flexibility and bargaining power over host country governments.

The second perspective further recognizes the importance of local complementary assets. Hennart (2009) argues that MNEs entering foreign markets to exploit their firm-specific advantages must bundle these advantages with local complementary assets. This implies that the entry modes used and what happens afterward would be simultaneously determined by MNEs and the owners of local complementary assets. This is similar to the revised internationalization process model, in which internationalization is theorized as an entrepreneurial process that emphasizes the identification of opportunities and taking advantage of these opportunities through learning of the host country market, creating new knowledge, and building relationships and trust with local partners (Johanson and Mattsson, 1987, 1988; Johanson and Vahlne, 2009). Of course, in this process, local firms can acquire knowledge and new technological abilities from MNEs.

From the perspective of this stream of literature, in studying Chinese firms' strategies in collaborating with Gulf nations firms, it is necessary to answer the following questions: (1) Do these Chinese firms have strong enough capabilities or superior resources to collaborate with their host country partners, or they are just offering low-cost labor; (2) Had these Chinese firms accumulated enough experience in the international market when they started to do business in the Persian Gulf region; (3) How important it is to make use of local complementary assets; and (4) Do these Chinese firms as latecomers in the international oil and gas industry have different strategies compared with well established IOCs.

The third stream of literature is about the key characteristics of Chinese SOEs (CNPC, Sinopec, and CNOOC are all SOEs, although they are also listed companies) (De Graaff, 2014; Ralston et al, 2006). This stream of literature is important because it is closely related to the strategies these Chinese firms could adopt in the Persian Gulf region. In fact, there are a lot of debates about these issues. For example, are these firms autonomous for-profit firms having their own capabilities and strategies or they are just vehicles of China's state-capitalism, threatening liberal values and interests (Andrews-Speed, 2004; Andrews-Speed and Dannreuther, 2011; Bremmer, 2009; Downs, 2007; Jiang, 2012; Kong, 2011).

Similarly, some studies find that Chinese SOEs are making impressive progress in technological innovation: the State Grid Corporation of China (SGCC) is now a leader in the world in developing and applying Ultra-high-voltage (UHV) electricity transmission technology, and the firm is playing a leading role in working with other members of the International Electrotechnical Commission (IEC) in setting up international standards in UHV technology (Liu, 2014); Datang Telecom Technology & Industry Group (Datang), the key developer of TD-SCDMA (time division synchronous code division multiple access), one of the 3G international technology standards for wireless communication, is also an SOE (Gao, 2014). However, there are studies that even argue that SOEs are not effective in technological innovation (Shen and Feng, 2010; Shi, 1998). What is the situation of the three giant SOEs in this study?

Research Methods and Data

This study is not about testing any specific existing theory, but rather is trying to develop some understanding of the key characteristics of these Chinese firms' strategies in collaborating with Gulf nations firms, so we use a case study method (Eisenhardt, 1989; Glaser and Strauss, 1967; Yin, 1989).

Research Setting

The paper is based on examining the practices of three Chinese firms doing business in the Persian Gulf region: CNPC, Sinopec, and CNOOC. CNPC is the largest producer and supplier of crude oil and natural gas, and a major producer and supplier of refined oil products and petrochemicals in China. CNPC is also one of the world's leading integrated energy companies, ranked 4th in the 2014 Fortune 500 companies. Sinopec is the second largest producer and supplier of crude oil and natural gas, and the largest producer and supplier of refined oil products and petrochemicals in China. Sinopec was ranked 3rd in the 2014 Fortune 500 companies. CNOOC is the largest offshore oil and gas producer in China and was ranked 79 in the 2014 Fortune Global 500.

These firms have been doing business in the Persian Gulf region for many years, expanding their collaborations with local firms, and becoming important players in this region. For example, CNPC began to explore business opportunities in this region in the early 1990s and have developed effective collaborations with firms in Iraq, Iran, Oman, Kuwait, etc.

Data Collection

Data directly related to Chinese firms' doing business in the Persian Gulf countries are collected using two methods. The first is through face-to-face interviews. As mentioned earlier, we adopted an interview-based case study method in this paper, so data collection is based mainly on interviewing people who are familiar with the three companies' practices of doing business in the Persian Gulf countries. The second method is using public data. This is necessary because the study covers many years and involved many people, so it's hard to rely only on an interview to collect data. It's also possible because there are rich public data in the form of news, company reports, and academic studies.

In addition, we also collected data indirectly related to Chinese firms' doing business in Persian Gulf countries. Among these data, those about China's energy structure and energy policy are important for us to develop a good understanding of the background of Chinese firms' doing business in Persian Gulf countries. For example, on the one hand, China is increasingly relying on importing oil to support its economic development (Table 1). On the other hand, in China's energy consumption, in 2012 oil and gas only accounted for about 24% (oil 19%, gas 5%), far less than coal, which accounted for about 67%.

In fact, according to the 2014-2020 action plan of the Chinese central government (Table 4), by 2020, coal consumption will decline but will continue to be the main energy, accounting about 62%, and oil and gas consumption will be kept at a relatively stable level of 23% (though oil will drop to 13%, and gas increase to 10%). Put differently, China will continue to stick to the policy of self-sufficient with regard to energy supply.

Table 4: China's Energy Mix in 2012 and 2020 (%)

	2012	2020
Coal	67	62
Oil	19	13
Gas	5	10
Other	9	15
Total	100	100

Source: composed by author

Data Analysis

Following the case study methodology, data analysis was conducted simultaneously with data collection (Glaser and Strauss, 1967; Strauss and Corbin, 2008). The research iterated between raw data, emerging concepts, related literature, and interrelationships among the concepts. A clear pattern emerged from the data analysis: Chinese firms are taking very high risks in doing business in the Persian Gulf region, and they have applied related strategies to manage these risks.

Key Findings of the Study

This part reports the key characteristics of strategies used by Chinese firms to do business in the Persian Gulf region: taking high risks, leverage existing capabilities, partnering with stakeholders, taking advantage of China's foreign policy, and taking advantage of the fast growing domestic economy.

Taking High Risks

The first characteristic of strategies used by Chinese firms is to take high risks in doing business in the Persian Gulf region, which has a lot of international relations, besides significant economic, social and political uncertainties. For example, CNPC was exploring opportunities of doing business in Iraq in 1994 and quickly gained trust from local partners. In 1996, CNPC actually signed an agreement with Iraq to develop the Al-Ahdab oil field.

However, it turned out that the UN sanctions on Iraq and the following US-led invasion made it impossible to develop this oil field.

CNPC did not give up its efforts in developing businesses in Iraq. In fact, it became the first international company to go to Iraq after the war: in November 2008, CNPC signed a development service contract with the Ministry of Oil of Iraq to develop the Al-Ahdab Oilfield. Following the Al-Ahdab Oilfield contract, CNPC partnered with British Petroleum (BP) and won a technical service contract with Iraq's State Oil Marketing Organization (SOMO) and South Oil Company (SOC) to expand production in the Rumaila Oilfield, the largest oilfield in Iraq and the 6th in the world, with oil reserves of about 17 billion barrels, on November 3, 2009. Partnered with Total, Petronas, and Iraq's Missan Oil, CNPC also signed a contract to increase oil production in the Halfaya oilfield, in January 2010.

It's worthwhile to note that the above-mentioned contracts are not at all lucrative in terms of making a profit. For example, in the Rumaila Oilfield contract, CNPC would be paid $2 per barrel oil produced. Many MNEs argued that it would be very hard to break even at this level of payment. This might be an important reason leading to the introduction of more favorable policies by the Iraq government in oil and gas exploration and development in order to attract more investment in later years.

Sinopec is also willing to take high risks in doing business in the Persian Gulf region. In fact, Sinopec's first international oil and gas exploration and development project is in Iran: in 2001 Sinopec signed a contract to develop the Kashan Block in the Central Iran Basin. This oilfield has highly complex geological conditions, and international companies conducting exploration activities before Sinopec failed to find oil in this area.

It seems that taking high risks in doing business in the Persian Gulf region is related to Chinese oil and gas firms being latecomers, compared with the giant IOCs, in the international market (Sampson, 1991; Troush, 1999). When IOCs have more choices, they do not have to take as high risks as Chinese firms. In fact, right after the war in Iraq, few IOCs were willing to take as low profit as CNPC (and BP) in order to win the bidding for oil and gas exploration and development.

Leverage Existing Capabilities

Taking high risks requires effectively managing and reducing the risks. One strategy used by Chinese oil and gas firms is to leverage their existing capabilities. For example, CNPC has accumulated strong capabilities in areas such as technological innovation and providing total solutions for oil and gas exploration and development.

In technological innovation, CNPC has developed advanced technologies in many areas, and the following are some examples: (1) non-marine petroleum geology theory, which is the theoretical basis for discovering the largest oilfield, the Daqing oilfield, in China, and other oilfields such as Liaohe, Xinjiang, and Huabei; (2) Technology for the development of large-scale, heterogeneous sandstone oilfields; (3) 3D seismic data acquisition and faster penetration drilling technology; and (4) seismic data processing and interpretation software.

For example, BGP is a leading geophysical service company in the world and a subsidiary of CNPC. In 2002, BGP's seismic data processing software suppliers decided to stop selling their software to BGP or imposed highly restrictive terms on BGP in the using of their software. This made it highly difficult if not impossible for BGP to expand in the international market. Supported by CNPC, BGP decided to develop its own seismic data processing and interpretation software in January 2013, and released its integrated seismic data processing and interpretation software in December 2004, GeoEast V1.0. The GeoEast software is now widely used in not only BGP but also other subsidiaries of CNPC.

A unique capability of CNPC is to provide a total solution for oil and gas exploration and development. In contrast to most IOCs, CNPC is a highly integrated company and has strong capabilities in the whole oil and gas value chain, including exploration, development, oil refining, and even financing. For example, in 2007 Changqing oilfield, a subsidiary of CNPC, was producing 21 million tons of oil equivalent. In 2008, Changqing set up the ambitious goal of reaching 50 million tons of oil equivalent by 2015 despite deep doubts from very influential experts outside the firm.

By leveraging the advantages of being a part of an integrated company, Changqing proactively initiated innovations to apply existing capabilities on a much larger scale, develop new capabilities, utilize external technologies at a much larger scale, help local suppliers to replace expensive foreign materials with locally produced ones, and apply new IT technologies. In 2012, the company reached 45 million tons of oil equivalent. It was also in 2012 that Changing surpassed Daqing and became the largest oil and gas producer in China. In 2014, Changing reached the production of 55 million tons of oil equivalent.

In addition, when CNPC started to do business in Iraq after the war, it had accumulated rich experience in international operations. For example, in 1993 and 1995, CNPC won the tenders to develop Block 7 and Block 6 of Peru's Talara Oilfield. Although the oil field is small and old, it has played a very important role in testing and refining CNPC's strategies of doing business in the international market, including applying proven technologies and providing integrated solutions for oilfield development.

Specifically, Talara is an almost abandoned oil field with a development history of more than 100 years. By applying its sophisticated theories and proven technologies, CNPC was able to not only drill high-yield new wells but also restore shut down wells. This led to the field's oil output increasing from 1,600 barrels per day when CNPC took over to more than 7,000 barrels per day.

CNPC was also able to replicate its success in Peru to other countries such as Venezuela, Ecuador, Kazakhstan, and Uzbekistan. For example, at the Caracoles and Intercampo Oilfields in Venezuela, CNPC increased the daily oil production from 4,900 barrels to 40,000 barrels by combing the restoration of the old oil wells and drilling new high-yield wells.

Applying the experience accumulated in other countries and capabilities developed in China, CNPC has been able to effectively carry out its exploration and development work and reduce risks, especially economic risks faced. For example, under the Al-Ahdab Oilfield contract signed in November 2008, CNPC was expected to reach the oil production level of 25,000 barrels/day within the first three years and 115,000 barrels/day by the end of the sixth year. In fact, CNPC was able to build up an annual capacity of 6 million tons by 2011.

Similar to CNPC, Sinopec is also active in leveraging their existing capabilities to reduce risks in doing business in the Persian Gulf region. For example, as mentioned earlier, Sinopec's first international oil and gas exploration and development project, Kashan Block, has highly complex geological conditions. By applying its advanced technologies and experiences (especially those accumulated in the Shengli oilfield), Sinopec was able to drill high yield wells in the Kashan Block. In fact, the success in the Kashan Block improved the host country's confidence and laid the foundation for Sinopec to win more exploration and development contracts in Iran, including the Garmsar oilfield and the Yadavaran oilfield.

Partnering with Stakeholders

Developing close partnerships with various stakeholders is another strategy used by Chinese oil and gas firms to reduce the high risks involved in doing business in the Persian Gulf region. One example is to partner with MNEs. This is also possible. On the one hand, these MNEs are also interested in doing business in the Persian Gulf region. On the other hand, these MNEs also want to reduce their risk of doing business in this region. For example, in developing the Rumaila Oilfield, CNPC partnered with BP, and in developing the Halfaya oilfield, partnered with Total and Petronas.

Partnering with MNEs in many cases is based on important collaborative relationships developed in the past. For example, BP began to do business in China in the early 1970s. BP was an important investor when PetroChina, CNPC's subsidiary, was listed in 2000. BP and CNPC also formed a joint venture (JV) in Guangdong province to develop the gas station business. In fact, by the end of 2014, 'CNPC had 35 joint exploration and development projects in operation, including 15 for crude oil, nine for natural gas, 10 for CBM and one for shale gas. In 2014, these projects produced 4.15 million tons of crude oil and 6.34 billion cubic meters of natural gas, which totaled 9.20 million tons of oil equivalent' (CNPC, 2014).

In addition to partnering with MNEs, developing strong partnerships that could make real contributions to the host country is another strategy. For example, CNPC believes that it is important to stick to the principle of "mutually beneficial cooperation for common development" and develop mutual trust with host country stakeholders such as the government, company partners, the local communities, NGOs, etc.

In Latin America, in order to protect the rain forest, CNPC has actively implemented a safe, green and resource-conserving development model, and is granted the 'Best HSE/ Sustainable Development Award'. In Sudan, CNPC has made an important contribution to not only the development of the oil and gas industry but also promoting local economic development through, for example, the development of the whole oil and gas value chain.

In the Persian Gulf countries, Chinese firms also followed the strategy of making as many contributions as possible to the host country. For example, in Iraq, CNPC has been paying a lot of attention to improving local employees' skills by appointing a tutor or co-worker to each local employee. Helping create job opportunities and subcontracting works to local firms as long as they could fulfill them (for example, campsite construction, equipment foundation works, bounding walls, roads, and other construction projects) are other practices to contribute to host country development.

Taking Advantage of China's Foreign Policy

As mentioned earlier, an important feature of China's foreign policy is to respect other countries' sovereignty and promote collaboration that benefits both sides. This has helped China to develop good relationships with many countries in the world. For example, after the war ended in 2000 in the Republic of Angola, China has been very active in developing mutually beneficial relationships between the two countries, including offering financial support to help with developing Angola's economy. Now the two countries have developed very close and effective economic collaborations. In fact, Angola has been China's second largest oil exporter country for several years (CNPC, 2014; Zhang, 2009), although its oil import to China was limited before 2000.

The three Chinese firms studied in this paper have clear understandings about and have been active in taking advantage of the favorable international relations created between China and other countries. CNPC's development in Sudan is a good example (Jin, 2012). In 1995, Sudan's President visited China and invited China to help develop its oil and gas industry. The Chinese government agreed and offered financial support. Because of this opportunity, CNPC has become a major player in the Sudan oil and gas industry. In fact, CNPC's construction of the 1506 km pipeline for transportation of oil from the Heglig oilfield to the Port Sudan within 11 months demonstrated CNPC's strong engineering capability and helped CNPC to win more contracts in the international market.

CNPC's development in the Central Asian countries such as Kazakhstan, Turkmenistan, and Uzbekistan is another example. Specifically, the Central Asia-China Gas Pipeline starts at the Turkmen-Uzbek border city Gedaim and runs through central Uzbekistan and southern Kazakhstan before reaching China, and the mutual trust and close collaboration between China and these countries are the bases for the effective construction and operation of this pipeline (Zheng, 2014).

Chinese firms doing business in the Persian Gulf region have also benefited from China's foreign policy. For example, China has been willing to offer political and diplomatic support to Iraq and Iran when it believes it is fair and appropriate. China's position and influence in international organizations such as the UN also make it possible for it to offer this kind of support (Troush, 1999).

Taking Advantage of the Fast Growing Domestic Economy

Taking advantage of the fast growing economy in China is also an important element in Chinese firms' collaboration with host country firms in the Persian Gulf. Specifically, collaboration is not limited to the host countries but has extended to China's domestic market. In addition to importing more oil and gas, Chinese firms have been exploring new approaches to collaborate with firms from the Persian Gulf region.

For example, NDRC approved a 50:50 JV between Sinopec and Kuwait Petroleum Corporation (KPC) in China in 2011. The JV planned to invest about 59 billion RMB Yuan to process 15 million tons of crude oil and produce 1 million tons of Ethylene. This project helps not only Sinopec to develop closer collaborations with KPC but also PKC to benefit directly from the rapidly developing Chinese market.

Conclusion

In this paper, we tried to develop some understanding of the key characteristics of strategies used by Chinese oil and gas firms (CNPC, Sinopec, and CNOOC) in their doing business in the Persian Gulf region. We found that these Chinese firms have to take very high risks in order to collaborate with host country firms in this region. Of course, they also have to develop strategies to lower the risks involved.

We did not look at future strategy choices for the Chinese firms given that there are some importance structural changes in the world oil and gas industry. For example, after 2000, the reliance on US oil imports from the Persian Gulf region has been declining steadily, while China has been importing more and more from this region. In fact, the US is relying less on importing energy from other countries, especially during the shale gas boom (Gold, 2014). The overall supply and demand for oil and gas in the international market seem to be more balanced, at least in the short run.

The international relationships in the Gulf region are also changing rapidly, and China is now able to develop stronger relationships with the Gulf nations. China and Iran decided to lift their relationships to "strategic partnership" during the meeting of the two countries' foreign ministers in Beijing in September 2015. In the same month, Mr. Zhang Yi, Chairman of the Chinese State-owned Assets Supervision & Administration Commission (SASAC), visited Iran, as the special envoy of the Chinese government. During the meeting between Mr. Zhang Yi and the Iranian First Vice President Eshaq Jahangiri, the First Vice President said that Iran welcomes Chinese President Xi Jinping's development strategy of "one road, one belt", and believes that Iran could play an important role in this strategy.

In fact, Mr. Zhang Yi actually brought with him 18 executives of large companies, including 12 Fortune 500 firms, to explore opportunities to collaborate in areas of industry, agriculture, transportation, etc. As mentioned previously, large SOEs in China have developed strong technologies capabilities. Accordingly, both Mr. Zhang Yi and Mr. Eshaq

Jahangiri believe that Chinese firms, including SOEs, could help with upgrading Iran's industrial capabilities and improving its infrastructures.

The US and Russia have developed deep strategic relationships over a long time covering security, energy, economics, etc, which is reflected in a dense "knowledge-ecology" between them and the Gulf. Compared with the US and Russia, China is a latecomer in the Gulf region. In fact, China is entering the world political and economic stage very fast, but both China and regions around the world have to very quickly develop knowledge and understanding of each other. This is also the case concerning Sino-Gulf relations. We plan to study these issues and their impact on Chinese firms' strategies in the Gulf states in the future.

References

Amsden, A. (2001) *The rise of "the rest": Challenges to the West from late-industrializing economies.* Oxford: Oxford University Press.

Andrews-Speed, Philip. (2004) *Energy Policy and Regulation in China.* The Hague: Kluwer Law International.

Andrews-Speed, Philip & Ronald Dannreuther. (2011) *China, Oil and Global Politics.* London: Routledge.

Bartlett, C.A. and Ghoshal, S. (2002) Managing across borders: The transnational solution. Harvard Business School Press.

Bower, T. (2009) Oil: Money, politics, and power in the 21st century. New York: Grand Central Publishing.

Buckley, P. J. & Casson, M. (1976) *The future of multinational corporation.* London: Macmillan.

Buckley, P. J. & Casson, M. (1988) A theory of cooperation in international business. In F. J. Contractor & P. Lorange, *Cooperative strategies in international business: 31-53.* Lexington, MA: Lexington Books.

Buckley, P.J. (1996) Government policy responses to strategic rent-seeking transnational corporations. *Transnational corporations* 5(2): 1-17.

Bremmer, I. (2009) State Capitalism Comes of Age. The End of the Free Market?. *Foreign Affairs* 88(3): 40-56.

Chandler A.D. and Redlich, F. (1961) Recent development in American business administration and their conceptualization. *Business History Review*, Spring 103-28.

CNPC, (2014). Annual report 2014.

Dannreuther, Ronald. (2011) China and Global Oil: Vulnerability and Opportunity. *International Affairs* 87(6): 1345-1364.

De Graaff, N. (2014) Global Networks and the Two Faces of Chinese National Oil Companies. *Perspectives on global development and technology* 13: 539-563.

Downs, E. (2007) China's Quest for Overseas Oil. *Far Eastern Economic Review*, (September 2007): 52-56.

Dunford, R., Palmer, I., and Benveniste, J. (2010) Business model replication for early and rapid internationalization: The ING direct experience. *Long Range Planning* 43: 655-674.

Dunning, J. H. (1992) Multinational enterprises and the global economy. Addison-Wesley: MASS.

Eisenhardt, K. M. (1989) Building theories from case study research. *Academy of Management Review*, 14(4): 532–550.

Gao, X. (2014) A latecomer's strategy to promote a technology standard: The case of Datang and TD-SCDMA. *Research Policy*, 43: 597-607.

Glaser, B.G., & Strauss, A.L. (1967) *The discovery of grounded theory.* NY: The Aldine Publishing Company.

Goladthau, A. and Witte, J. eds. (2010) *Global energy governance: The new rules of the game.* Brookings Institution Press and Global Public Policy Institute.

Gold, R. (2014) *The boom: how fracking ignited the American energy revolution and changed the world.* New York: Simon & Schuster Paperbacks.

Hennart, J.-F. (2009) Down with MNE-centric theories! Market entry and expansion as the bundling of MNE and local assets. *Journal of International Business Studies* 40:1432-1454.

Hymer, S. H. (1976) The international operations of national firms: A study of direct foreign investment. MIT Press, Cambridge: MA.

Jin. D. (2012) The Study on the Developing History of China Petroleum Company in Sudan. *Journal of Xi'an Shiyou University (Social Science Edition)*, 21: 63-69.

Jiang, B. (2012) China National Petroleum Corporation (cnpc): A Balancing Act Between Enterprise and Government. In: David G. Victor, David R. Hults & Mark C. Thurber (eds.), *Oil and Governance. State-owned Enterprise and the World Energy Supply*. Cambridge: Cambridge University Press.

Johanson, J. and Mattsson, L.G. (1987) Interorganizational relations in industrial systems: a network approach compared with the transaction cost approach. *International studies of Management and Organization*, 17(1): 34-48.

Johanson, J. and Mattsson, L.G. (1988) Internationalization in industrial systems: a network approach. In N. Hood and J.E. Vahlne (Eds), *Strategies in global competition*: 468-486. London: Croom Helm.

Johanson, J. and Vahlne, J-E. (2009) The Uppsala internationalization process model revisited: From liability of foreignness to liability of outsidership. *Journal of International Business Studies* 40: 1411-1431.

Kong, B. (2011) 'Governing China's Energy in the Context of Global Governance.' *Global Policy* (2): 51-65.

Liu, J. (2014) *A Study on indigenous technology innovation in UHV transmission*. Tsinghua University School of Economics and Management master thesis.

Liu, Z., Sun, X., Gao, X., & Hu, J. (2006) *Cases of technological innovation at CNPC*. Beijing: The Oil Industry Press (in Chinese).

Meyer, K.E., Yen, T.T. (2006) Market penetration and acquisition strategies for emerging economy. *Long Range Planning* 39: 177-197.

Price-Smith, A. (2015) *Oil, illiberalism, and war: An analysis of energy and US foreign policy*. The MIT Press.

Ralston, D., Terpstra-Tong, J., Terpstra , R., Wang, X., and Ergi, C. (2006) Today's state-owned enterprises of China: Are they dying dinosaurs or dynamic dynamos? *Strategic Management Journal* 27: 825-843.

Sampson, A. (1991) *The seven sisters: The great oil companies and the world they shaped*. New York: Bantam Books.

Shen, Q. and Feng, K. (2010) From production capacity to technological capability: an institutional and organisational perspective, *International Journal of Technology Management* 51(2, 3, 4), 258-281.

Strauss, A., & Corbin, J. (2008) *Basics of qualitative research techniques and procedures for developing grounded theory* (2nd ed.). London: Sage Publications.

Troush, S. (1999) *China's Changing Oil Strategy and its Foreign Policy Implications*. Center for Northeast Asian Policy Studies, the Brookings Institution.

Yetiv, S. (2011) *The petroleum triangle: oil, globalization, and terror*. Cornell University press.

Yin, R. K. (1989) *Case Study Research: Design and Methods*. Newbury Park, CA: Sage Publications.

Zhang, K. (2009) The composition analsysis of China's oil import sources since 2000. *Petroleum and Petrochemicals Today* 17(6): 17-21, 44.

Zheng, H. (2014) *The Management of projectized long-distance -cross-border Pipeline Company under complicated Circumstances*. Tsinghua University MBA Dissertation.

10

An Empirical Test of the Predictors of National-Expatriate Knowledge Transfer and the Development of Sustainable Human Capital

Sheikha Shamma bint Sultan bin Khalifa Al Nahyan
and Laura L. Matherly

Introduction

For the last several decades, it has been increasingly recognised that people are an organisation's most valuable asset in achieving a sustainable competitive advantage (Kiessling and Harvey, 2006). Professionals and managers who have a significant depth of specialised knowledge can add value to the organisation. By transferring knowledge internally through interactions among people and tasks, organisations can build a competitive advantage (Argote and Ingram, 2000). In 1981, the GCC was established in Abu Dhabi as an economic and political intergovernmental alliance among Bahrain, Kuwait, Oman, Qatar, Saudi Arabia and the UAE. Together, these six states possess 40% of the world's proven oil reserves; and for the last two decades, they have sought to invest their substantial oil rents to improve their human capital and global competitiveness (MENA Report, 2012; World Bank, 2014).

Today these nations are embarking on major economic diversification projects with the goal of transitioning from an economy fueled by natural resources to one that is innovative, knowledge-based, competitive and self-sustaining. To support rapid economic development, especially in the private sector, a growing expatriate population has provided a range of professional skills based on fixed term employment contracts. The total number of expatriates is the GCC is estimated at over 48.1 million (Global Labour Markets Migration, 2015). At the high end of the spectrum, the percent of expatriates in the UAE and Qatar is estimated to be just over 80% while in Saudi Arabia, it is about 30%. The UAE not only attracts the most expatriates, it has the highest retention rate, i.e., 88% of the expatriates residing in the UAE desire to remain in the country (Gulf Talent, 2015).

An important driving force that has relevance to building a knowledgeable workforce in the GCC is that population growth is expected to continue to be strong, especially in the cities, creating a large pool of labour, the majority of which will be 25 years of age or under (MENA Report, 2012). Based on the foregoing, a significant challenge to countries in the Arabian Gulf is to ensure that qualified nationals with relevant skill sets are available to build the industries and sustain the economic development that is desired for long-term stability and growth. While knowledge management is important to all organisations, the employment of a large number of non-nationals represents an opportunity to create and capture knowledge that resides in individuals that is easily lost due to turnover. Therefore, it is important to determine the factors that promote and encourage knowledge transfer to develop a knowledgeable native workforce so that organisations can reap the benefits of employing expatriate professionals. Knowledge exchange and building capabilities is a strategic priority. Leadership is required to encourage a culture and structure that supports learning, focuses on meeting and exceeding organisational goals, in order to not only build the right competencies in the workforce but lead to successful development of local economies.

This study tests a theoretical model that identifies the key factors underlying the successful use of workplace quotas for nationals to develop human capital and therefore, build the performance of a work force that creates a sustainable competitive advantage in the global arena. The premise of this paper is that localisation is not successful when a national is hired into a position and an organisation meets the strategic targets for nationalisation. Nationalisation is successful when the local workforce has the expertise and capability to first, sustain the organisation and more importantly, lead the country's economic development to meet their strategic goal of being one of the highest ranked economies in the world competitiveness. Therefore, sustainability is contingent on effective and efficient knowledge transfer between targeted nationals and expatriates who possess specialised expertise.

Theoretical Model

In general, knowledge transfer is viewed by practitioners and scholars as important to building an organisation's competitive position especially in the context of a global, interconnected knowledge-based economy. It is an important element in the operations of organisations especially in their efforts to ensure the effective development of competencies and skills and the sustenance of a compelling competitive advantage (Hocking et al., 2007). A significant degree of evidence points to the capability for organisations to transfer knowledge between units as one of the most essential contributing factors to superior organisational performance (Zanini and Musante, 2013).

Moreover, managing professionals is complex for a number of reasons. Generally, the knowledge that is most valuable, i.e., is a source of competitive advantage, is tacit in nature (McDonnell et al., 2010). Experienced, contractual professionals may occupy strategic positions and perform key functions that are central to the organisation's operations and so they are required to exercise discretion and decision making based on specialised knowledge (Jones, 1983).

The theoretical model shown in Figure 1 depicts the proposed relationships being explored in this study. In brief, strategic enablers are hypothesised to affect knowledge transfer: transparency, trust, incentives, motivation and interactions between nationals and non-nationals. First, whether leadership has established a transparent culture with high levels of trust will influence knowledge sharing. The human resource management infrastructure and incentives are hypothesised to play an important role in knowledge transfer and the development of employee competencies. These strategy enablers are hypothesised to predict knowledge transfer. Next, the national/non-national exchange of knowledge is predicted to be a function of both dyad members' qualifications, motivation and receptivity. The final two factors that are important to successful nationalisation efforts have to do with whether knowledge transfer is linked to process improvement and the attainment of organisational goals. To this end, important key performance indicators must be identified with appropriate targets that are linked to nationals and non-nationals performance. Specifically, to achieve world class performance levels and the implementation of best practices, process improvements that are linked to meeting and exceeding customer expectations is required. These factors working in concert will result in developing the future generation of leadership that is capable of sustaining the economic growth and prosperity of the country.

Figure 1: Knowledge Transfer Model

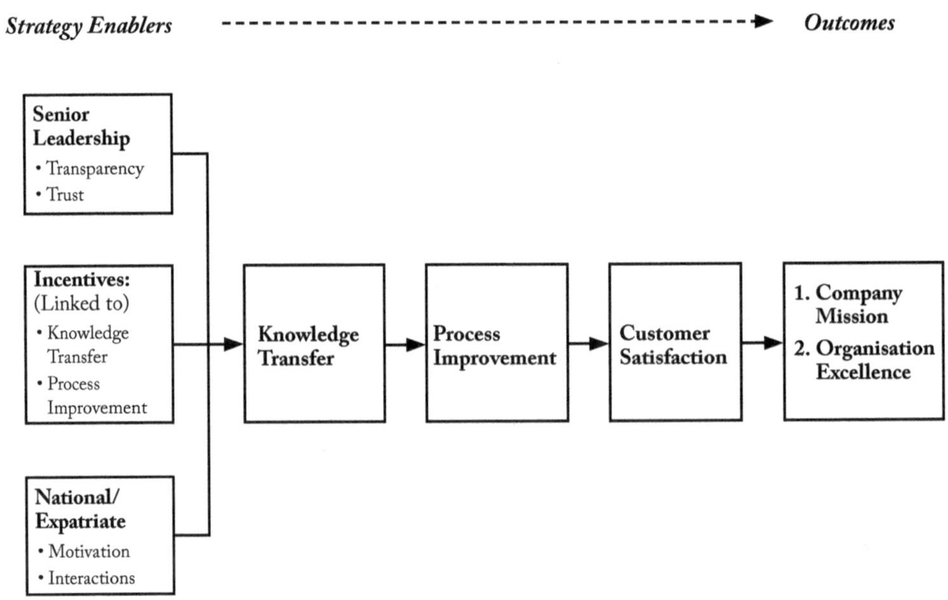

Senior Leadership – Transparency and Trust

Building a learning organisation is impacted by the interplay of many factors, and when senior leadership takes the initiative to create the conditions that will encourage knowledge transfer and learning, they are executing a strategy that will position the organisation for success. Leadership affirms the legitimacy for knowledge exchange and provides direction to the firm. In the proposed model, effective knowledge management of national/non-national learning requires leadership to establish transparency and build trust through encouraging the flow of information throughout the organisation. In a global, knowledge based economy, secrecy and control of information is rapidly being replaced by transparency. According to Werbach (2009), transparency is a mind-set and "the sharing of information above and beyond what is needed is an act of trust that will engage your constituents and enlist them on your behalf." The role and influence of senior management in organisational culture is widely acknowledged (Kor and Mesko, 2013). This includes influencing transparency about decisions and information as well as employees' sense of trust in the company.

In financial reporting and corporate governance, transparency is regarded as beneficial to the organisation as it ensures accountability in management (Hood and Heald, 2006). Timely and reliable disclosure of financial information has been widely recognised as essential to shareholders' and stakeholders' ability to make informed decisions. More recently, transparency has evolved to encompass a more comprehensive meaning to include openness, accountability, credibility and trust applicable throughout the organisation.

Organisations that embrace transparency as a core management strategy will foster an environment and culture of openness. At the same time, the process of creating a transparent system and culture in the organisation can be time-consuming; therefore, and in the face of competing priorities, such as production and operations, managers may place their attention on urgent day-to-day priorities at the expense of important but less urgent priorities, such as transparency. Transparency includes the articulation and communication of the decision-making rationale, i.e., the basis for a decision, such as promotions, rewards and incentives, and how they relate to business results.

It is incumbent upon senior management to develop ways of promoting transparency to close information gaps within an organisation (McNaughton, 2004). By agreeing on the major principles that underlie workforce quotas, managers can create a common agenda among employees in an organisation. If a single member of the top management is not fully involved in the information chain, then the employees under that individual are likely to be out of sync with the rest of the organisation. This can lead to suboptimisation, lower productivity and an overall under performance of the entire company (Studer, 2009).

Organisational transparency is important to successful workforce localisation because it promotes a clear and direct way to link incentives to desired outcomes or behaviours. Employees are more likely to share the perception that the organisation intends for them to have and be in sync with the organisation's objectives, mission, vision and goals. In

order to achieve this, top management must be united and like-minded and consistently communicate the same message (Neyland, 2007). By widely sharing information in the organisation, management will gain the trust of employees and encourage learning. In addition, whether management encourages open and critical review of work processes with a view toward learning and improvement will affect the level of knowledge transfer.
Trust in the leadership of an organisation is an important element in an organisation pursuing continuous improvement in performance and increasing competitiveness through knowledge management (Thompson et al., 2009). According to Bodnarczuk (2008), "trust is the foundation of all human interactions, and the cornerstone upon which high-performing organisational cultures are built" (include page number). Yeow and Blazjewski (2007) examined the impact that a trustful organisational culture has on knowledge transfer within multinational organisations. A culture of trust results in greater employee willingness and ability to share knowledge among themselves, within and also across subsidiaries. Their findings point to the need for organisations and managers to foster a supportive culture involving openness and trust as bases of communication. The leaders of an organisation, i.e., management, play a crucial role in shaping a knowledge transfer culture and this can be implemented when employees are treated fairly within the organisation. Based on the foregoing, the following is hypothesised:

H1: There is a positive relationship between transparency and knowledge transfer.

Transparency is defined as the extent to which management explains the reasons for key decisions and is open about sharing information. This refers to the free flow of information from top-level managers to the front line and vice versa. Knowledge transfer refers to how frequently nationals and expatriates share their experience and help each other improve their skills on the job including critical thinking and problem-solving skills.

H2: There is a positive relationship between trust and knowledge transfer.

Trust is measured by whether the company keeps its promises and is fair. Trust is also assessed by the extent to which expatriates trust nationals and whether nationals trust expatriates.

Rewards

Knowledge transfer depends whether the organisation incentivises and rewards learning especially, management's support of collaboration and open communication (Subramanian and Soh, 2009). Based on Pucik's (1988) work with strategic alliances, there are two types of learning capacity: the actual reward system and a clear learning agenda that is communicated to relevant parties. Reward systems can be reflected in career advancement, performance appraisals and other forms of formal and informal recognition. In a study of 3 types of learning capacity – resource based, incentive-based and cognitive based – Simonin (2004) found that incentive-based or incentives that are clear between the reward system and sharing information and a well communicated learning agenda, had the most significant effect across samples. Without rewards and incentives, expatriates are not likely to devote attention to building other employees' talent, especially if they are rewarded for work results only. If incentives are based on country of origin, i.e., national or non-national, knowledge transfer will not be valued, effective or efficient.

> H3: There is a positive relationship between incentives linked to knowledge transfer and knowledge transfer.

Incentives linked to knowledge transfer concern the extent to which each group, expatriates or nationals, are recognised and rewarded for sharing knowledge, e.g., in their performance appraisal.

> H4: There is a positive relationship between incentives linked to process improvement and knowledge transfer.

The definition of incentives linked to process improvement addresses whether incentives are based on performance, not "who you know" and specifically, whether rewards are linked to improvements in efficiency and effectiveness.

National/Expatriate Motivation and Interactions

Although organisations, and especially leadership, hold the responsibility for developing effective cultures, nationals and expatriates can play a critical role in ensuring the establishment of an efficient environment for knowledge transfer. Transferring knowledge is an intentional behavior and therefore, the motivation and willingness to transfer and seek opportunities to develop others is a key component to transfer knowledge. The behaviours of both the national and the expatriate need to be aligned with an objective of working closely together to ensure knowledge sharing. Providing an employee a competitive salary and benefits does not assure that he/she will be motivated to perform well; however, an organisation can determine what motivates that individual and strategically manage individual expectancies regarding the link between behaviours and desired outcomes (Vroom, 1964).

Creating a progressive and learning organisation is essential to the process of knowledge-sharing and transfer. Self-determination theory offers a framework for understanding intrinsic motivation for sharing knowledge (Deci and Ryan, 2000). Autonomous motivation (intrinsic motivation) is when an individual shares knowledge for intrinsic ends e.g., an individual enjoys sharing knowledge with his/her peers. Externally regulated motivation is a type of controlled motivation that is based on a reward and/or punishment system and the implementation of deadlines (Cabrera and Cabrera, 2002). Autonomous/ intrinsic motivation leads to better results and is related to the concept of learning organisation (Gagne and Deci, 2005). In this type of environment, motivation originates from inside and therefore, the sharing of knowledge becomes important in and of itself, independent of external rewards. Self-motivated people go the extra mile and have personal stakes in sharing of knowledge beyond the reward system operating in controlled motivation framework.

The nature of the expatriate-national relationship is set within the context of government policy regarding labour laws, which can affect job security and therefore, motivation. A policy in the GCC is the Kefala or sponsorship system that monitors expatriates working within the region. The policy dates back to the 1950s and regulates the association between employers or companies and their foreign workers (Murray, 2012). This system has an important economic objective to offer temporary and rotating employment that could be rapidly exploited in booming periods and reduced during less prosperous periods. Job conditions, salaries and benefits tend to favour national employees. Expatriates have less job security because they are on short-term contracts. In addition, government policies requiring national workforce quotas could have a negative impact in the workplace where expatriate workers are reluctant to transfer knowledge to their national counterparts in order to protect their jobs. Therefore, expatriates may not be motivated to transfer knowledge that ultimately could result in losing their job.

Knowledge transfer is driven by communication and information flows vertically and horizontally, between individuals, as well as within and across groups throughout the organisation (Alavi and Leidner, 2001). Since the same individual will have different degrees of knowledge transfer with others, depending on factors such as ability, motivation, and trust the appropriate level of analysis is the individual dyad. Frequent dialogue and communication created by common experiences among individuals facilitates the transfer of tacit knowledge, e.g., knowledge that is not explicit and recorded in documents or manuals (Nonaka and Takeushi, 1995). Interactions are necessary to learning by experience and building a common understanding or mindset among employees. Therefore, if nationals and non-nationals work physically close to each other and interact frequently together, there will be more knowledge transfer. While interactions provide the participants exposure to information, nationals and non-nationals may differ in the extent to which they are willing and able to receive or provide knowledge. Nonetheless, interactions are a necessary prerequisite to transferring knowledge. Therefore, the following propositions are hypothesised:

H5: There is a positive relationship between national and expatriate motivation and knowledge transfer.

Motivation is operationalised by asking respondents to indicate whether they actively seek out opportunities to help each other develop; whether they are willing to share knowledge with each other; and if they encourage others to see them as "approachable".

> H6: There is a positive relationship between interactions between nationals and non-nationals and knowledge transfer.

Interactions concern how close a respondent's workstation is to the other group, e.g., national or expatriate, and the frequency of interaction on a formal and informal basis.

Knowledge Transfer and Process Improvement

External professionals are capable of encouraging knowledge transfer because of their ability to bring a significant degree of knowledge and experience to host countries and companies (Hocking et al., 2007). Knowledge and expertise that emanates from work in a variety of countries and business scenarios, enhances their level of credibility and trust from recipients (Simone and Leon, 2012).

Expatriates and nationals may transfer knowledge that supports the firm's ability to meet important organisational goals. Learning from these collaborative experiences will not only benefit the organisation but will build the work-related competencies of both groups. Knowledge that is central to the organisation's performance will be desired and acquired to the extent that there is a performance reward system linking knowledge transfer to goal attainment. The more knowledge transfer that occurs between nationals and non-nationals that is work-related, the more likely that internal organisational processes will be improved.

> H7: There is a positive relationship between knowledge transfer and process improvement.

Process improvement encompasses whether nationals and expatriates work together to improve work processes and specifically, if improvements in organisational performance and key performance indicators are the result of knowledge sharing between the groups.

Process Improvement and Customer Satisfaction

Consumers evaluate and judge a firm and its products based on performance in a variety of parameters, such as quality, dependability, cost, timeliness of service, corporate communications, and, of course, reputation. Organisations that meet and exceed customer expectations by improving internal processes can expect revenue growth and an increased market share. Fulfilment of these factors influences consumers' purchase decisions, which, in turn, determines an organisation's competitiveness and overall financial performance.

Furthermore, these aspects, which comprise an organisation's value propositions to the consumers, directly relate to their internal processes.

> H8: There is a positive relationship between process improvement and customer satisfaction.

Customer satisfaction assesses the extent to which nationals and expatriate relationships and exchange of information improve customer satisfaction.

Method

The questionnaire consisted of just over one hundred questions. The survey primarily used a five-point Likert-scale format with response categories ranging from strongly disagree to strongly agree. Scales were reverse coded where appropriate. Based on a pilot test, the survey took about 20 minutes to complete. After obtaining approval from their management, 32 graduate students at Zayed University on an Executive MBA course distributed the surveys as part of an action research project. The surveys were distributed across a range of management levels as well as employees in departments where both nationals and expatriates worked. A cover letter was included that requested voluntary participation and assured the respondents of confidentiality. Surveys were administered electronically and manually and resulted in 496 useable surveys from 65 companies and the overall response rate was 43%.

Measures

The instruments for the variables in the study were developed from a review of the relevant literature and qualitative interviews with subject matter experts. A pilot study was conducted to pretest for content validity. The final survey measure used a response format ranging from 1 = strongly agree to 5 = strongly disagree. To ensure that the measures were empirically distinct and establish both convergent and divergent validity between the scales, a principal factor analysis was used to develop the subscales. Based on a principal component analysis, the number of survey items was reduced from 61 to 48. All factors had eigenvalues higher than 1.0. The resulting measures and survey items are reported in Appendix 1. The means, standard deviations, coefficient alphas and correlations for the variables are shown in Table 1.

Transparency was assessed with 6 items that addressed the free flow of information from managers to employees and vice versa. Moreover, respondents were asked to indicate whether management was open in explaining key decisions and communicating bad news. To assess trust, 5 items asked whether the company kept its promises, was honest and could be trusted to do the right thing. In addition, whether management was fair and whether nationals and non-nationals trusted each other was asked. The coefficient alphas were .87 and .84 for transparency and trust, respectively.

Incentives were partitioned into two measures – 5 items that were linked to knowledge transfer and 2 items that were linked to process improvement. The former asked participants to indicate the extent to which they were recognised or rewarded for sharing knowledge each other. Specifically, non-nationals were asked whether they were rewarded for sharing knowledge with nationals and nationals were asked whether they were rewarded for sharing knowledge with non-nationals. Similarly, both groups were asked if they contributed to the other's development. The two items that measured links to process improvement concerned whether incentives were based on performance (not who you know) and improvements in efficiency and effectiveness. The coefficient alphas were .73 for linked to knowledge transfer and .90 for linked to process improvement.

Motivation was addressed with 6 items that measured the extent to which respondents actively sought out opportunities to share knowledge, be seen as approachable and help each other develop. The coefficient alpha was .90. Three items measured interactions and asked respondents whether their workstation was physically close others and the frequency of formal and informal interactions. The coefficient alpha was .73.

Knowledge transfer items assessed the frequency with which the two groups shared experience and knowledge and helped each other improve critical thinking, problem solving and skills. The 6 items had a coefficient alpha of .90.

The following demographic variables were controlled for in the statistical analysis: gender, years experience, position (senior manager, middle manager, and manager), nationality, education level, and monthly income.

Table 1: Correlation Matrix

	Mean	Standard Deviation	Coefficient Alpha	Transparency	Trust	Rewards-Know Transfer	Rewards-Improvement	Motivation	Interactions	Knowledge Transfer	Process Improvement	Customer Satisfaction	Nationlization Success
Transparency	2,67	,75	,87	1,00									
Trust	2,40	,69	,84	.69**	1,00								
Rewards-Know Transfer	2,83	,81	,73	.60**	.52**	1,00							
Rewards-Improvement	2,56	,88	,90	.56**	.56**	.58**	1,00						
Motivation	2,02	,66	,90	.34**	.41**	.20**	.27**	1,00					
Interactions	2,20	,74	,73	,03	,07	,06	,08	.29**	1,00				
Knowledge Transfer	2,40	,47	,90	.78**	.78**	.73**	.62**	.64**	.33**	1,00			
Process Improvement	2,25	,64	,82	.41**	.44**	.45**	.40**	.49**	.31**	.75**	1,00		
Customer Satisfaction	2,10	,72	,68	.44**	.51**	.32**	.37**	.50**	.32**	.64**	.57**	1,00	
Nationlization Success	2,28	,79	,82	.55**	.51**	.53**	.43**	.25**	,01	.56**	.35**	.35**	1,00

496 sample size *p< .05; **p<.01

Results

A total of 496 surveys were completed which included 38% female and 62% male respondents. The average age reported was 41.4 years old. The majority of the respondents were nationals at 52%, followed by Arab and Asian expatriates both at 19% and Western expatriates at 8%. A substantial proportion of the participants were managers, i.e., 30%, where 11% were senior managers, 15% were middle managers and 14% were managers. For those respondents who were not managers, most indicated that they were in a specialist, technical or professional position. The high proportion of specialists or professionals in the sample was confirmed by analysing the income levels of respondents in this category, which showed that 71% made 30,000 dirham a month or higher, i.e., >$8,000 or >5000£. The majority of the respondents,

86%, had a bachelor's degree or above. A total of 48 organisations were included in the study. Approximately half (49%) of the respondents were in the public sector, whilst only 16% worked in the private sector and 35% work in semi-private entities.

Analysis

The hypotheses were tested with an ordinary least squares regression analysis. The results are shown in Table 1. None of the control variables were significant indicating that knowledge transfer did not vary by gender, years experience, position, nationality, education level or monthly income.

Table 2: Regression Analysis

Variables	Standardized Coefficient	P-Value
Intercept	.00	
Gender	.01	.5187
Years Experience	.00	.7876
Position	.00	.7030
Nationality	-.02	.1323
Education Level	.00	.8044
Monthly Income	.01	.6199
Transparency	.26	.0000
Trust	.27	.0000
Rewards-Knowledge Transfer	.34	.0000
Rewards-Process Improvement	.02	.0926
Motivation	.31	.0000
Interactions	.19	.0000

Based on the results in Table 1, Hypotheses 1 and 2 were supported. The correlation between transparency and knowledge transfer was strong, $r = .78$ and between trust and knowledge transfer, $r = .78$, both at $p \leq .01$. Hypotheses 2 and 3 are supported by the correlation analysis. Incentives that are linked to knowledge transfer and process improvement are associated with higher levels of knowledge transfer, $r = .73$ and $.62$ ($p \leq .01$), respectively. Both national and expatriate motivation are significantly associated with knowledge transfer indicating support for Hypothesis 5, $r = .64$, $p \leq .01$. Hypothesis 6, which concerned the relationship between formal and informal interactions between nationals and expatriates, was supported in the correlation analysis where $r = .33$, $p \leq .01$. The correlation between knowledge transfer and process improvement was significant, $r = .75$, $p \leq .01$ which supports Hypothesis 7. Last, high levels of process improvement are associated with high levels of customer satisfaction as hypothesised in Hypothesis 8, $r = .57$, $p \leq .01$.

Table 2 presents the results of the regression analysis used to simultaneously assess the relationship between all the independent variables and knowledge transfer. Multiple and stepwise regression are used to develop the final model so that the smallest number of independent variables that predicted the majority of the variance in the dependent variable were selected. In Table 2, based on the standardized regression coefficients, the strongest predictor of knowledge transfer is when rewards are linked to knowledge transfer followed by motivation, trust, transparency and interactions which were significant at $p \leq .001$. The analysis of the final regression model with the independent variables that are significant is shown in Table 3. The F statistic is significant ($p > .0001$) with a multiple R of .97 and adjusted R-square = .95.

Table 3: Final Regression Analysis

Variables	Standardized Coefficient	P-Value
Intercept	.00	
Transparency	.17	.0000
Trust	.19	.0000
Rewards-Knowledge Transfer	.21	.0000
Motivation	.22	.0000
Interactions	.12	.0000

Discussion and Conclusions

The model for thriving in the modern, global economy can only be accomplished through the development of human capital so that the right employee capabilities are available to pursue whatever strategy is required to compete, respond to changes and take advantage of emerging opportunities. Developing human capital is an ongoing and long-range endeavor that when integrated with process improvement will lead to superior firm performance and competitive standing. The extent of knowledge transfer between nationals and non-nationals is important to organisations that desire to build knowledge from non-nationals before their contract ends.

The dynamics of knowledge transfer between nationals and non-nationals centre around not only the motivation and interaction of the individuals, but also the leadership and reward structure of the organisation. By focusing their talents, knowledge and skills on managing the intra and interdepartmental complexities that operate as barriers to cost improvement, efficiency and growth, leaders can create the environment and systems necessary to improve profits and revenue growth (Ray et al., 2003). The results of this study indicate that the degree of knowledge transfer is not related to any of the control variables, i.e., education levels, income levels, gender, years experience, income or nationality. In other words, a culture of transparency and trust where expatriates and nationals have the opportunity to interact as well as the motivation and rewards to interact will result in effective knowledge sharing. This relationship held across demographics including education, income, gender and nationality. Overall, motivation to transfer knowledge was very high in the sample group, which may indicate that in positive, motivating environments, it is possible to facilitate knowledge sharing by increasing interactions and rewards for knowledge sharing. Future research based on groups that have lower levels of overall motivation, may ascertain if these relationships hold under different conditions. Similarly, since the UAE has experienced a period of strong economic growth and prosperity and attracts/retains more expatriates than other GCC countries (GulfTalent, 2015), additional research is needed to determine if the predictors of knowledge transfer hold in situations that are less favourable.

While motivation had a strong relationship with knowledge transfer, the relationship between motivation and rewards was significant but low indicating that there are other factors influencing motivation. Future research should investigate how to increase motivation and include, for example, the impact of job security and short-term employment contracts on the motivation to transfer knowledge. Relatedly, if expatriate employment contracts are more similar to national contracts, does knowledge transfer improve? What parts of the employment contracts are most important in predicting knowledge transfer, e.g., salary, tenure, benefits, etc.

Learning organisations are skilled at systematic problem solving, experimentation, learning from their experiences and transferring knowledge (Garvin, 1993). The role of organisational leaders is to create the conditions, i.e., trust, transparency and incentives, for knowledge transfer to take place. When leaders share information and create a culture where information flows openly, knowledge transfer is more likely to occur. Similarly, knowledge transfer is encouraged when management builds trust by actions that demonstrate fairness and consistent follow through.

An effective nationalisation program that builds competencies and closes developmental gaps requires individual development plans and goals that become part of the key performance indicators in a performance management system. While investing in training and development is important to learning some knowledge and skills, the acquisition of tacit knowledge occurs primarily on the job. Management of nationalisation programs should identify the critically important knowledge that must be transferred between nationals and expatriates and evaluate the success of the program with measureable goals that are tied to knowledge transfer and achieving business results or the targets of the organisation.

There are few studies that examine how organisation acquire new knowledge from expatriates and facilitate the transfer of knowledge among nationals and expatriates. While this study included a large sample of organisations and respondents, a limitation is that the sample was comprised largely of professionals in public or semi public organisations. This affects the generalizability of the results and more research is needed in private organisations including all employee/organisational levels. For example, are the relationships that predict knowledge transfer in this study generalizable to nonprofessional employee samples?

This study adds to the extant knowledge on understanding the most important predictors of knowledge transfer between expatriates and nationals and underscores the importance of senior leadership in establishing a culture of transparency and trust. Similarly, linking rewards and incentives to knowledge transfer and process improvement will lead to the desired improvements not only in internal processes, but ultimately, in customer satisfaction. Last, the motivation of both nationals and non-nationals, as well as the opportunity to interact, will foster knowledge transfer. The implications of this study clearly delineate the important building blocks to capitalise on learning through the exchange of knowledge between expatriates and nationals.

References

Alavi, M. & Leidner, DE 2001, 'Review: Knowledge management and knowledge management systems: Conceptual foundations and research issues'. *MIS Quarterly*, 25(1), 107-136.

Argote, L. & Ingram, P. 2000, 'Knowledge transfer: A basis for competitive advantage in firms'. *Organizational Behavior and Human Decision Processes*, 82(1), 150-169.

Bodnarczuk, M. *The Organizational Trust Index as window into organizational culture*. Boulder, CO: Breckenridge Institute.

Cabrera, A. & EF Cabrera 2002, 'Knowledge-Sharing Dilemmas'. in *Organization Studies* 23 (5): 687-710.

Deci, EL & Ryan, RM 2000, 'The "What" And "Why" Of Goal Pursuits: Human Needs And The Self-Determination Of Behavior'. Psychological Inquiry 11 (4): 227-268. doi:10.1207/s15327965pli1104_01.

DeTienne, KB, Dyer, G., Hoopes, C. & Harris, S. 2004, 'Toward a model of effective knowledge management and directions for future research: Culture, leadership, and CKOs'. *Journal of Leadership and Organizational Studies*, 10(4), 26-43.

Gagne, M. and Deci, EL 2005, 'Self-Determination Theory And Work Motivation'. *Journal of Organizational Behavior* 26 (4): 331-362.

Garvin, DA 1993, '*Building a learning organization*'. Harvard Business Review, July-August, 78-91.

Global Labour Markets and Migration. 2015. 'GCC: Total population and percentage of nationals and non nationals in GCC countries (latest national statistics, 2010-2015)'.

Gulf Talent. 2015, '*Employment and Salary Trends in the Gulf*': www.gulftalent.com

Hocking, J. B., Brown, M. & Harzing, AW 2007. "Balancing global and local strategic contexts: Expatriate knowledge transfer, applications and learning within a transnational organization", *Human Resource Management*. 46(4): 513-533.

Hood, C. and Heald, D. 2006. "*Transparency: The Key to Better Governance?*", Oxford University Press, Oxford.

Jones, GR 1983, '*Transaction costs, property, rights and organizational culture: An exchange perspective*', *Administrative Science Quarterly*, Vol. 28, No. 3, Organizational Culture, pp. 454-467.

Kor, YY. & Mesko, Y 2013, 'Dynamic managerial capabilities: Configuration and orchestration of top executives' capabilities and the firm's dominant logic', in *Strategic Management Journal*, 34 (2) (2013) 233-244.

McDonnell, A., Gunnigle, P. & Lavelle, J. *2010*, '*Organizational learning in multinational companies: Explaining inter-organization variations*', *Human Resource Management Journal*, 20(1) 23-43.

McNaughton, D. 2004. '*The Pros & Cons of Organizational Transparency*'. Retrieved April 25, 2014 from: http://organizationevolution.com/the-pros-cons-of-organizational-transparency/

Murray, HE 2012. 'Hope for reform springs eternal: How the sponsorship system, domestic laws and traditional customs fail to protect migrant domestic workers in GCC countries.' *Cornell International Law Journal, 45*, 461-485.

Neyland, D. 2007, 'Achieving transparency: The visible, invisible and divisible in academic accountability networks,' *Organization.* 14(4): 499-516.

Nonaka, I., & Takeuchi, H 1995, '*The Knowledge-Creating Company: How Japanese Companies Create the Dynamics of Innovation*'. Oxford, UK: Oxford University Press.

Pucik, V. 1988, 'Strategic alliances, organizational learning, and competitive advantage: The HRM agenda,' *Human Resource Management*, 27: 77-93.

Ray, G., Barney, JB and Muhanna, WA 2003, 'Capabilities, Business Processes, and Competitive Advantage: Choosing the Dependent Variable in Empirical Tests of the Resource-Based View', *Strategic Management Journal*, 25(1) 23-37.

Studer, Q. 2009, '*Straight A Leadership: Alignment, Action, Accountability.*' Gulf Breeze, FL: Fire Starter Publishing.

Simone, P., and Leon, C. P. 2012. "Knowledge is Power? An Inquiry into Knowledge Management, Its Effects on Individuals Creativity and the Moderating Role of an Entrepreneurial Mindset". *Academy of Strategic Management Journal,* 11(1): 43-57.

Simonin, BL 2004, 'An empirical investigation of the process of knowledge transfer in international strategic alliances', *Journal of International Business Studies* 35.5 (2004): 407–427.

Subramanian, A. M. and Soh, PH 2009, 'Contributing knowledge to knowledge repositories: Dual role of inducement and opportunity factors'. *Information Resources Management Journal*, 22(1): 45- 62.

Thompson, MP & Jensen, RJ & DeTienne, K. 2009. 'Engaging embedded information: Toward a holistic theory of knowledge transfer in organizations', Competitiveness review: an international business journal, 19(4), 323-341.

Vroom, V.H. 1964. *Work and motivation*. NewYork: Wiley.

Werbach, A. 2009, 'Strategy for Sustainability: A Business Manifesto'. Harvard Business Press.

World Bank, 2014. World Development Indicators. World Bank, Washington. DC. Available at: http://tinyurl.com/gme64kl

Yeow, P. & Blazjeski, S. 2007, 'Knowledge and Innovation: Opportunities and Challenges to Knowledge between Multinationals and their Subsidiaries (running head: Knowledge Transfer in MNCs)' in *Kent Business School.*

Zanini, MT & Musante, M. 2013, 'Trust in the Knowledge Economy', *Journal of Business & Industrial Marketing*. 6(28): 487-493.

11

Interprofessional Healthcare Collaboration: A Qualitative Exploration of the Facilitators and Barriers to Effective Collaboration in Qatar[1]

Michael K. Corman and Jason Hickey

Introduction

Qatar is a small, peninsular country in the Persian Gulf. Traditionally, its economy was based on trading among nomadic Bedouin tribes as well as fishing and pearl farming undertaken by a small population of coastal Arabs. However, the economy and the country itself have undergone a rapid change since large reserves of oil and natural gas were discovered in 1939. Despite its size, Qatar is now an extremely wealthy country; it has the highest number of millionaires per capita ("The economy: The haves and the have-nots", 2013) and is ranked as having the highest GDP per capita in the world (Qatar National Bank, 2014).

The newly acquired wealth and associated development have not come without challenges. Since 1984 the population has grown from just under 350,000 to over 2.2 million in January, 2015 (Qatar Statistics Authority, 2015). Air pollution is among the highest in the world (World Health Organization [WHO], 2014), many feel that social and cultural values are being eroded (Fellow, 2013), and there have been growing concerns about the mistreatment of foreign workers (Al Jazeera, 2014; Grant and Falconer, 2013). Health has also been negatively affected. Over 75% of men and women over 19 are overweight or obese (Ng et al., 2014), major depressive disorder has become the main cause of disability adjusted life years (i.e., DALYs) and ischemic heart disease is the number two cause of death after road injury (Institute for Health Metrics and Evaluation, 2013).

A substantial amount of Qatar's financial resources are being utilized to address these health challenges. In fact, healthcare in Qatar is undergoing a period of major reform driven by the strong economy and a vision for a world-class healthcare system. Transnational knowledge relations and foreign healthcare professionals play a large role in building this healthcare system. Qatar's National Health Strategy outlines the framework that will be used to develop a world class healthcare system (General Secretariat for Development Planning, 2008). According to the National Health Strategy (General Secretariat for Development Planning, 2008: 56), interprofessional education has the potential to improve the "cohesiveness and connectedness" of the national healthcare system; it is listed as a tool for improving service delivery for a more integrated system of healthcare (Supreme Council of Health, 2011).

Progress has been made in recent years towards increasing the profile and implementation of more effective interprofessionalism in Qatar. Some examples include the formation of the Qatar Interprofessional Health Council, a recent research grant from the Qatar National Research Fund awarded to develop and implement an undergraduate IPE training program, the formation of the Qatar Academic Health System (QAHS), and funding provided by the QAHS to develop a post-licensure IPE training project. The study described in this article was conducted as a part of the post-licensure interprofessional education (IPE) project.

Background

The WHO (2010) recently published a report on interprofessional education and collaborative practice (IPECP). They define IPE as occurring "when two or more professionals learn about, from and with each other to enable effective collaboration and improve healthcare outcomes" (World Health Organization, 2010: 13). Collaborative practice, referred to in this article as interprofessional collaboration (IPC), "occurs when multiple health care workers from different professional backgrounds provide comprehensive services by working with patients, their families, carers and communities to deliver the highest quality of care across settings" (WHO, 2010: 13). Authors of this WHO (2010) report assert that there is now enough evidence to conclude that IPECP can lead to strengthened health care systems and improved health outcomes. They also describes mechanisms that can affect IPECP, such as staff training, managerial commitment, communication strategies, and built environment, many of which can act as both barriers and facilitators.

One limitation of the IPECP model is that is does not include a feedback loop from IPC to IPE. Clark (2011), for example, describes how an investigation of barriers and facilitators for IPC led to recommendations for improved IPE. Jones and Delaney (2014) interviewed mental health professionals about their understanding of collaborative practice. The findings, that collaboration needs to be contextualized and that practitioners need shared philosophical understandings, have obvious implications for IPE. Similarly, the authors of a study that explored conceptual differences towards collaboration at the bedside conclude that "unless such complexities and contextual factors are addressed in the preparation for collaboration it will continue to fall short" (Barrow et al., 2014: 119). These studies demonstrate that an exploration of collaborative practice on the front lines can feed back into, and inform, development of IPE initiatives.

While the WHO's (2010) report was based on evidence from around the globe, very little input was included from the Middle East due to the relative absence of IPECP literature from the region; little is known about the regional "complexities and contextual factors" that affect IPECP. Unique to Qatar and others areas in the region is that the majority of the workforce is made up of "expats" from around the world. For instance, according to the Qatar Statistics Authority, in 2012 roughly 82,000 Qataris were employed in Qatar versus 1.25 million non-Qataris (Qatar Statistics Authority, 2015). Although public statistics are unavailable that describe the diversity of the healthcare workforce in Qatar, we can extrapolate from these statistics (and our personal experiences) that the healthcare workforce is diverse. This creates a unique context whereby healthcare workers come from multi-cultural and multi-lingual backgrounds with varied educational training.

The current article aims to contribute to the international discourse on IPECP by sharing results from a qualitative study conducted in Qatar. Facilitators and barriers to IPC are presented and discussed within the context of a health care system undergoing transformation. The results provide a new perspective on IPECP from a rapidly developing Arabic country.

Methods

This study was a part of a larger interprofessional education initiative to provide interprofessional workshops for licensed practitioners. Ten healthcare professionals who work in Qatar were interviewed using semi-structured, open-ended interview questions, for an average length of 30 minutes (range from 16-48 minutes). According to Gill, Stewart, Treasure, and Chadwick (2008: 292), this interview length is common in healthcare research, with healthcare interviews lasting, "on average", between "20-60 minutes". Participants were selected through a convenience sample of workshop participants. Participants represented a range of disciplines, including nursing, pharmacy, medicine, dietetics and nutrition, and respiratory therapy. The purpose of the interviews was to explore different facilitators and barriers to interprofessional collaboration based on the lived experience of participants. We also aimed to shed light on some of the work strategies healthcare workers use to activate interprofessional collaboration in practice. Hence, while some of the interviews were relatively short in length, it is important to emphasize that the goal of these interviews was not to provide "statistical representativeness . . . but instead the chance to look in detail at how selected people experience the world" (Brinkmann, 2013: 59) in relation to the topics at hand.

It is important to mention that this convenience sample is a privileged sample, despite being participants working on the front lines of healthcare in Qatar. Hence, a limitation of this research is that it focuses on healthcare workers in mid-level administrative/

managerial/educational positions. However, this sample provides a perspective from those most involved in planning and implementing change in Qatar's healthcare system. While this view is certainly important, the perspectives of those working predominantly at the bedside would add another important dimension to this work. Another limitation is that we focused on tertiary care and did not include participants from primary care facilities. As such, future research could add to the findings offered in this article by investigating a more representative sample consisting of participants from multiple specialties.

Through an exploration of the facilitators and barriers to interprofessional collaboration, we have three main goals. First, we provide an understanding of interprofessional collaboration in a cultural context – Qatar specifically, and the Arab region more broadly – that to date has garnered little attention in the IPE/IPC literature. Second, we provide a complex and diverse understanding of the barriers and facilitators that structure the opportunity to work together collaboratively and interprofessionally. Third, by shedding light on the barriers and facilitators to interprofessional collaboration in practice, we aim to inform and better facilitate interprofessional educational strategies and practices in Qatar specifically, and the Middle East more broadly.

This study was organized by a qualitative description design as described by Sandelowski (2000). According to Sandelowski (2000: 336), qualitative description studies are more focused on describing a social phenomenon and are "not highly interpretive" or theoretically informed[2] compared to more traditional qualitative research designs (e.g. phenomenology, grounded theory, and ethnography). In other words, "Although no description is free of interpretation . . . qualitative description, as opposed to, for example, phenomenological or grounded theory description, entails a kind of interpretation that is low-inference, or likely to result in easier consensus among researchers" (Sandelowski, 2000: 335). As such, the goal of the design deployed here is to offer a summative description of events based on the "everyday terms of those events" described by participants (Sandelowski, 2000: 336).

Furthermore, qualitative description studies often can take on "hues" or "tones" of other approaches. In the case of this study, there are distinct hues of phenomenology (Creswell, 2002; Moerer-Urdahl and Creswell, 2004) and ethnography[3] (DeVault and McCoy, 2006). More specifically, the phenomenological hue can be seen in the interview questions aimed at exploring the lived-experiences of healthcare workers in relation to working collaboratively with other professionals, and the ethnographic hue in the questions aimed at focusing on what people actually do to collaborate on the front lines of health care. Phenomenological questions asked participants to discuss different facilitators and barriers to interprofessional education based on their lived-experiences working in Qatar. The ethnographic interview questions asked participants to discuss the following questions: How is working collaboratively with other healthcare professions accomplished? Similarly, what kind of work is involved in working together? What is some of the behind-the-scenes, "backstage" or negotiative work (Reeves, 2011: 17) involved in accomplishing interprofessional collaboration? With this combination of questions, the overall goal of these interviews was to "allow interviewees to express their opinions and ideas in their own words" (Esterberg, 2002: 87) on the topic of IPC.

Interviews were transcribed in their entirety. Data were organized into themes through an open-coding process followed by a more focused coding process. All coding was performed by both researchers. In order to increase descriptive validity, initial coding was compared between researchers until consistency was achieved. Analysis was facilitated by the use of NVivo 10, a qualitative data analysis software. Because the first author is a sociologist and the second author a nurse, this reflexive and interactive process provided us with an interprofessional perspective on the data. Furthermore, Reeves (2011: 9) explains that few studies on IPC "employ a wider sociological lens" (see also Thistlewaite et al., 2011), so the interprofessional design of this research also facilitated a more generous and nuanced perspective on the data. Through collaboration, complex themes were identified, which are reported in the findings section below.

Ethical review of this study was granted by the Conjoint Health Research Ethics Board at the University of Calgary and by Qatar's Joint Institutional Review Board. All participants provided written informed consent. In order to protect the anonymity and confidentiality of participants, we use pseudonyms below when we reference quotes and we refrain from identifying the nationality and occupation of participants.

Findings

The findings section is separated into three sections. The first and second sections report on the barriers and facilitators, respectively, to interprofessional collaboration as identified by participants. The third section reports on some of the informal strategies that participants deployed in practice in order to activate interprofessional collaboration. It is important to note that while we present these sections as separate and the themes within each section as being discrete, the sections and themes discussed should be seen as interfacing and, as a result, the themes run together at times. In other words, the themes we discuss in each section, and the sections themselves, are not necessarily separate from one another. Rather, we would like readers to view the findings reported below as overlapping and being recursively related to one another in diverse and complex ways that require further investigation.

Barriers

In this section we report on some of the barriers – factors/elements that impeded – to interprofessional collaborative practice. Participants identified a variety of factors that acted as barriers to working together. We grouped these barriers into four themes, including hierarchy, communication, diversity of education background of workforce, and time. We discuss each theme below.

Hierarchy

Four themes related to barriers to interprofessional collaboration were identified through the analysis, including hierarchy (organizational and medical), communication, the diversity of the workforce, and time. Hierarchy, whereby there is an imbalance in power between or within professions, was one barrier reported by participants. According to one participant, Asma, this barrier was, in part, the result of a work organization that privileged medical doctors over other healthcare providers, where "the overall control [is] from the medicine." Another participant, Rachman, explained:

> The biggest barrier is the hierarchy system whereby it's very much, like, the doctors are in charge of everything. They say what we need to do and that. Even the nurses think that way: 'It's not my job, it's the doctors who make the decisions.' So they are the leaders . . . Like the doctors wouldn't accept something that easily if a nurse suggests it. Like when we're doing education and activities, the doctors are happy to give us education but they will not accept [it] from us. Sometimes, simple things, like documentation, if we know more about [something] . . . they feel uncomfortable. 'Like who are we to tell them?'

Some participants explained that this hierarchy between professions led to a "siloed" workplace where, as one participant said, "no interaction between the physiotherapist and the nurse or the physician" (Amna). This was emphasized by participants who explained that some "old school people" (Sara) have preconceived stereotypical notions about other professions. Asma alludes to how these preconceived notions might be rooted in cultural differences of some healthcare workers. For example, she explained that, "some doctors, they will come from different countries and they have this idea or this assumption about the role of the nurse. That you are nothing. 'I know what you do. You do this, this, this, and you are here to help me.'" It is important to note that Asma mentioned how this siloed workplace was not just on the front lines but "sometimes, even the higher management" are "kind of siloed up on the top".

Where this hierarchy was in place, it seemed to be based on assumptions about other occupational groups and/or a lack of understanding of "what everybody has to do" (Asma). This lack of understanding can be viewed as part and parcel to how some participants discussed how some occupational groups in some of the workplaces were siloed from each other. Asma, for instance, mentioned that because there is this perception, "each discipline thinks that he's doing the best; no need for him to discuss with others".

Communication

As a result of being "siloed" from other professional groups, participants identified communication as another barrier to IPC. According to Rachman:

> Well, it's going to be one-way communication most of the time, right? And even if you want to ask for something, it's going to mean, like, "Please?" – as a favour – rather than that we're both equals where we can do that. We're not always equal. Of course, they have, I mean, I'm not downgrading their position as physicians, but we have our value and sometimes the physicians think of us as just people who take vital signs, or don't think for themselves yet.

This one-way communication can structure social interactions in ways where what some people say "goes" – is authoritatively listened to – while others are made voiceless – "it's pretty impossible to talk" (Noon). As Noon went on to explain, this one-way communication connects to policies and practices where some workers have to "just do" what they are told or else they can get into trouble; "I have to [just do it], or it will come down on me". With no professional body in place in Qatar to protect or offer support to workers who face this socially structured dilemma, professional discretion and concerns of professional practices have the potential to be subsumed or displaced by barriers to interprofessional collaboration in health care. Comparing her Western country of origin with working in Qatar, Noon explained the following:

> If you really thought you were gonna harm the patient by doing something ... Um, we could say [in her country of origin], 'No. I'm absolutely not doing that . . . That goes against everything I've learned, and that's dangerous for the patient, and I refuse to do that.' So you're allowed to do that ... we have colleges that protect us ... We don't have them here yet [in Qatar].

In a similar light, it appeared that some of the work settings of participants were characterized by a limited discourse of critical dialogue and questioning that prevented some individuals from speaking up and speaking out, hence acting as a barrier to communication. Noon exemplified this when she explained:

> It's a very social environment here. It's very pride-driven. You know, you don't, you don't, even if you don't think someone's doing a very good job, you don't really tell them that to their face, especially if you're a local to a local.

Diversity of Educational Background of Workforce

Participants also discussed how the diversity of a workforce can create an environment characterised by a "lack of trust . . . between disciplines" because the distinction between different occupational groups is "not clear for most of the people" (Asma). This uncertainty reinforces and likely contributes to a hierarchical or siloed workplace. Concomitant with a diverse workforce is different types of educational training and cultural differences that resulted in some workers having "assumptions" (Asma) of other workers, which further connects to the barrier of hierarchy discussed above. Asma explained this as follows:

> Because we come from different countries, we come from different education backgrounds, we come from different ways of seeing healthcare professions. For example, in my country nursing is prestigious, medicine is a pride. In some neighboring countries nursing is undermined or is not that prestigious. So if you are working with a doctor who comes from country X for example, the way they look to the nurse is different to my country, the clash will arise.

Another participant explained how her Western education taught her to "question" and "challenge" other healthcare workers (Amna). Amna contrasted this education with the educational backgrounds of many of the workers in Qatar, who, in her opinion, lacked this "educational exposure". In fact Amna explained that some of the training of workers "means they're subservient to the physician, and it's a massive mind shift to try and get them to challenge a practice, never mind engage in learning with other professionals".

It is important to note that a diverse workforce is not a barrier in and of itself. In fact, one could imagine many strengths that would come from having a diverse workforce, especially in healthcare, simply because patients themselves are diverse and come from a variety of socio-cultural-linguistic backgrounds. However in light of work environments that are characterized by the aforementioned barriers, especially hierarchical work settings that contribute to some healthcare workers being "othered", it is no surprise that diversity can act as a barrier to interprofessional collaboration.

Time

The barriers discussed above also connect inevitably to resource allocation in the hospital environment. In the current sample, this was particularly evident in relation to time, or the lack thereof, to collaborate interprofessionally. For example, some participants drew attention to how the time available to collaborate interprofessionally was not present in their work settings. Rachman explained the following:

> Time. When you don't have enough time to sit down and talk and discuss things, could be an issue. And when it's too busy - too many emergencies happening everywhere – and there's no time to talk together with the patient. Even when they're doing those rounds in the morning, for example, you cannot always get the lab and the pharmacist and everybody around to you. So that's an issue.

Another participant explained that while healthcare workers "would gain massively from engaging with others", some workers are not provided the time to do so; "they don't get allowed time to go and learn. They have to go on their own time, so it needs to be on weekends, or evenings. They don't get time back" (Amna). Amna went on to described how she finds it challenging to engage with some occupational groups in collaborative ways because "they are far too busy to spare the time during their work to come together as professionals to discuss". She then went on to explain how some work settings have time "built-in" to their work that facilitates "meeting on a regular basis" (discussed further in the next section).

While lack of time is an issue in engaging collaboratively, Asma expressed how this issue of time connects to other barriers, like hierarchy, discussed above.

> The doctors . . . they don't have too much time to be with you, or to work with each other. Some of them, they are willing to, but . . . each one is completely according to his mindset, how he's prepared to work with each other. I know some people, they don't like to attend even the lectures if there will be other [occupational groups] other than them.

Perhaps unique to Qatar, some participants discussed how under-staffing and the turnover of health care workers, an issue connected to resource allocation, the diversity of the workforce, and the political economy of the region, could impede interprofessional collaboration. One participant, for example, explained that even when interprofessional education programs are put in place, staff cannot attend because of "understaffing" (Nicol). Similarly, one participant expressed that staff turnover and lengthy HR processes made it a challenge to collaborate interprofessionally because "they're constantly having new people coming through" and "if someone quits . . . It takes a year at least to get most people hired . . ." (Noon).

Facilitators

While participants were able to speak about barriers to interprofessional collaboration in Qatar, they were also able to speak about facilitators to interprofessional collaboration. Many of these facilitators exemplified what is working well in terms of interprofessional collaboration. However, it seemed that the facilitators were fewer than the barriers participants spoke of. This difference suggests that more facilitators need to be put in place to support interprofessional collaboration in Qatar. Of the facilitators participants drew our attention to, structural opportunities, communication, scope of practice, and training were the four main themes identified.

Organizational/Structural Opportunities

Participants mentioned a variety of what we have labeled as organizational/structural opportunities that, in their opinion, facilitate interprofessional collaboration. For example, one participant spoke about an external group of healthcare practitioners being brought into their department for consultation and to improve work practices. This external group provided different education and training resources, assisted in the implementation of policies and practices, and provided an example of how to work collaboratively. As one participant explained, "So they [this external group] are doing a lot of education, and also trying to encourage work, working together, and working together in teams . . . so they're huge" (Noon). As such, this external healthcare group provided a structural opportunity that facilitated IPC by providing new education and training opportunities that aligned with the principles of IPE.

Other structural opportunities included allocating space that allows for collaboration and various workshops for participants to attend with other healthcare workers. In regards to the former, Sara said that having "the availability of small commons rooms or offices would facilitate [collaboration] as well." It is worth noting that based on observations made by the second author in his capacity as a clinical instructor, break rooms in hospitals where he has worked are currently siloed in a way similar to practice (as described previously). For example, nurses have one room to break in and doctors have another. Aside from a cafeteria, which is also open to the public, there is no common space where health professionals can interact informally. As such, having a physically integrated workplace setting could facilitate informal interaction between professionals.

Lastly, some participants spoke about how part of their work is "structured around meeting on a regular basis" (Amna) with other healthcare workers. In other words, structural opportunities, like set times to meet as outlined in daily or weekly schedules, and "paper rounds" where a group of health care workers would "sit together as a team with whoever is involved" (Sara) facilitated interprofessional collaboration. Essentially, when time was "built in to coming together" and work is "structured around meeting on a regular basis" (Amna), this facilitated interprofessional collaboration is facilitated. Amna further went on to explain:

> For example, every week we know to set time that the whole team will come together and just go through the patients' cases . . . It's planned for in everybody's schedule that at such time, on such day, we meet together and discuss the case load.

Communication and Training

Linked to the structural opportunities provided by/embedded within the work place setting, and in contrast to communication as a barrier discussed above, some participants discussed how communication between different professionals was facilitated when "everyone know[s] his role" (Tam). This created an environment characterized by mutual respect; "They are so co-operative. When we ask them, when I collaborate with pharmacy, with the anesthesiology, with obstetrics, they are co-operative" (Nicol). It is important to note that one participant who spoke about knowing the roles of other individuals outside of their occupational group worked in a smaller work setting. Amna explained:

> I think purely it's because [the hospital where I work] is a smaller area, and it's much more like everybody knows everybody . . . Everybody knows everybody, and everybody says hello. [Other hospitals in Qatar] have a much, much bigger workforce, and people don't take the time to get to know each other. They haven't got the time, really, because their process is quicker . . . Yes, we do have some very [busy] areas, but the staff have been there for a long time and maybe rotated from other areas, as well, so everybody knows everybody else.

Collaborative communication was also facilitated by training. Some participants explained that the people they work with are "well educated" (Raghda) and "have been taught communication skills . . ." (Sara). Furthermore, and in contrast to the barrier of medical hierarchy and communication discussed above, Sara explained that some of the individuals she interacts with are much more open to be challenged:

> 'Challenge me, tell me I'm wrong,' and then the next day we'll discuss it all together. It would make his day if I tell him, as a junior [health care worker], 'doctor, I think you're wrong. I've got some evidence that backs me up' . . . It'll make his day.

While many participants did not express this openness to critique in health care settings in Qatar, it is important to note that embedded in this participant's talk above is a suggestion that change is on the horizon. For example, educational upgrading is currently being undertaken in Qatar for the healthcare workforce. In fact, a mission of the University of Calgary in Qatar is to provide current nurses working in Qatar with an additional two years of university education so that they qualify for a Bachelor degree in nursing. In our interviews, it became clear that certain level of training and education had the potential to interface with and confront barriers to interprofessional collaboration, such as hierarchy and a lack of communication.

Connected to this discussion of openness to collaborate and critique is trust between healthcare workers, which one participant discussed in relation to the importance of education. Simply put, "Education is very important for all disciplines . . . if they trust each other, then they will like to work with each . . . [which] can lead to a better outcome, rather than working alone" (Asma). In addition to fostering trust, Sara explained how knowing what other people do, as learned through education and experience, creates an empathetic and "understanding staff." This participant exemplifies a theme that we heard from participants in our interprofessional education workshop; by the end of the workshop, it became clear that one way to help facilitate interprofessional collaboration is to provide educational programs that foster trust between different occupational groups.

Scope of Practice

Participants talked about how their scope of practice required individuals to work together as another facilitator to IPC. Examples of this included when a pharmacist collaborated with a nurse to check the medications of a patient or different clinicians working together for a "long stay" (Amna) patient requiring rehabilitation services. As one participant explained, because of institutional practices that require one group to consult with another group of healthcare professionals, "most of the time they need our recommendation to work" (Tam). While the need for referrals often required in hospital settings facilitates collaboration, the quality of this collaboration requires further investigation.

Work Strategies

We contend that in addition to the discussion of the barriers and facilitators above, an examination of some of the work strategies of participants in light of and in response to the aforementioned barriers and facilitators sheds further light on how IPC is activated in practice. Participants, for instance, mentioned a variety of work strategies that can be thought of as both facilitators of and responses to the barriers of interprofessional collaboration as discussed above. Such work strategies included building rapport with their colleagues, negotiation work, and modifying their language depending on the context.

It is important to note at the outset of this discussion that the strategies discussed below are likely not the only factor at work here. Certain elements of "social capital" (see for example Kritsotakis and Gamarnikow, 2004) linked to the participants being in a relatively privileged position (e.g. primarily located in mid-level administrative/managerial/educational positions) in their respective organizations probably interfaced with these work strategies. In other words, elements of privilege likely structured their ability to navigate and move beyond some of the barriers in place in their work settings. In discussing the concept of social capital and privilege, we seek to move away from an individualizing understanding of work strategies to a broader, more diverse and complex understanding of how work is structured by a multiplicity of social factors. For example, the convenience sample, as discussed above, consisted of individuals who were mostly in administrative/managerial/educational positions, had, for the most part, many years of experience, and, for roughly half of the sample, were educated and/or recruited from the West. The social capital of this sample would likely be very different from other front-line workers who hold less administrative and supervisory positions. Future research on this topic would significantly add to our discussion below.

Use of Language

Some participants described making modifications to how they spoke to certain individuals in order to collaboratively work together; they learned to organize their talk in ways that are responsive to their audience. For example, in talking to an individual who is higher up on the medical hierarchy, Rachman explained, "Sometimes what we do is we kind of, like, they're the boss but we are helping. Like, 'if it's okay with you?' There are a lot of politics. It also depends on the physician. Some physicians are fairly open. Some physicians are easily approachable." Another participant, Noon, expressed similar sentiments:

> Generally it would be the senior consultants, and even of that, it would be certain senior consultants . . . Maybe I would be a bit more forceful with my thoughts on a pat[ient]. . . on someone who's more junior . . . because they know me . . . but some of them are more willing to discuss it and tell you why they think something rather than saying, 'No. I said so, and that's what you do.'

Sara explained how she uses certain "buzzwords" in order to activate the interests of different professionals she needs to collaborate with:

> So, with the speech and swallowing therapist, you've got to say, 'Her swallowing might be unsafe.' That's it. That's your buzzword. If you tell her, 'she's got diabetes, hypertension, and she has a brain hypoxia, she's a transfer from ICU,' they're not interested . . . The neurologist, they're not interested about the unsafe swallowing. You tell them, 'She's got a brain bleed, and she's got thrombosis. She's got a blood clot. So we're going to give her a blood thinner that might worsen the brain bleed.' So that's the buzzword. You give them the full history, more or less, but you've got to use your buzzwords . . . You don't lie, you shouldn't lie . . . It's different from one specialty to another. For the nurses, the buzzword is, 'You haven't given this medication. The consultant is coming.'

Similarly, Noon explained how she had to change the way she spoke to some of the individuals she worked with in order to meet the needs of the context (e.g. cultural differences between healthcare workers) of her work setting:

> I had to take some time. I had to even, like, change the way I talked, because it, like I heard from other people, like my staff . . . I told them, like, 'Tell me what the feedback that you're hearing,' you know . . . That's the other thing. In [her country of origin], if someone doesn't like it, they're just telling you. But here, not. It's very, you know, talking about you behind your back, sort of. And, so I heard from my colleagues that, or my staff like, 'They don't like your voice.

> They think you're yelling at them.' What can I do? But I had to really, I did. I consciously tried to change my voice to be quieter.

The contrast in talk described above is telling. Whereas some of the participants modify their talk in ways to activate the interests of certain specialty groups (e.g. through the use of "buzzwords"), others activate the work of healthcare workers by structuring their language in a way that is organized by a medical hierarchy. Sara, who herself is located further up the medical hierarchy, mentioned how she uses a sort of boss-worker or bourgeoisie-proletariat trope to activate the work of nurses – "The consultant is coming." Similarly, when one participant mentioned how "there's lots of politics", this discussion of language use above exemplifies how "politics" (hierarchy) enters into and organizes how individuals orient and talk to other professionals.

Negotiation Work

Connected to our discussion of language use above, negotiation work was another strategy some participants used to activate collaborative work with others. For example, Al-Juan explained the following:

> So in some cases we are telling them we have some conflict or curiosity regarding the medication because this patient used to take this medication and you are sending out [a] prescription like this. Is there any reason? 'I'm sorry doctor ... but there is something in the pharmacy we have to be sure about it before giving medication to the patient ...'

The language use of this participant exemplifies a sort of subtle negotiation work that likely interfaces with some of the work strategies already discussed, such as using a specific way of speaking to the practitioner who is higher up on the medical hierarchy. Another participant, Tam, mentioned how she uses "the scientific base" to "negotiate" with other professionals. While this is one strategy that this participant deployed, she pointed out that such negotiation work is not always successful; "even [when] you negotiate with them or [tell] them the scientific base ... they [don't accept your opinion.]"

Rapport Development – Setting the Scene for Interprofessional Collaboration

Other participants discussed how they set the scene for interprofessional collaboration by building relationships of trust when they first began to work in Qatar. Noon explained how she began to build bridges of trust:

> [I would] always be willing to offer help, to do other things. You know, helping them turn the patients, or helping them reposition. Like, just saying to them,

> 'If you have any questions, if you don't understanding something.' So I would go around and kinda ask them like to, have [you] any questions about what we were doing.

Noon went on to explain that once this rapport was in place, it was easier to make her voice heard: "They've learned that they can trust me, and I know what I'm talking about. So if I have a meeting, say about nursing, they . . . say, 'Yeah, yeah. That's good. Do, do whatever.' No. I definitely had to earn my [respect]." Another participant, Borat, expressed similar sentiments in how he develops rapport. He explained that he "convinces" other occupational groups of the "benefits" of the knowledge his occupational group brings to patient care.

Discussion/Conclusion

This article reports on findings from a qualitative study that uses insights from phenomenology and ethnography to explore interprofessional collaboration in Qatar, where many health care professionals are foreign. We specifically focused on providing a complex understanding of the different barriers and facilitators to interprofessional collaboration and some of the different work strategies involved with accomplishing interprofessional collaboration on the front lines of healthcare. In doing so, our goal was two fold. We first sought to provide insights into interprofessional collaboration in a country that, to date, has limited if any research on the topic at hand. Second, by better understanding the different facilitators and barriers to IPC, and the different work strategies involved in accomplishing collaborative work, we can improve interprofessional educational strategies in Qatar by addressing some of the structural barriers discussed above and build off of what is working well – the facilitators – in Qatar.

One of the analytical insights this article draws attention to is that there are inequities in power leading to hierarchies in the workplace of participants. In other words, hierarchy, or lack thereof, seemed to be an overarching theme of our analysis that structures interprofessional collaboration either in positive or negative ways. While this hierarchy is sometimes necessary in healthcare, it can impede interprofessional collaboration, and it can also put healthcare workers in precarious positions with potential consequences to the quality of care provided to patients. Noon, for example, contrasted how in the country where she originally worked, she would have no problem questioning a doctor's decision. In fact, she explained that a doctor would typically "welcome the challenge from me . . ." In her current workplace, if she does not abide by the doctor's directives, "it will come down on me." Sara expressed a similar concern where, "If the consultant made a decision that I thought is not right – because I've read something, or I've come across something, or I've worked with something else, I'm not allowed to challenge them." In these examples, both participants might be forced to go against their professional judgement because of a system that structures what they can and cannot do. This can have consequences. As Borat explained, when collaboration between professionals is not present, "Sometimes we are missing data about a patient", which could seriously impact the quality of care a patient receives.

Our discussion points to the need for a complex examination of barriers and facilitators to interprofessional collaboration and caution for a one-dimensional individualistic view of such barriers and facilitators; it would be simplistic to assume, for example, that the educational or "cultural" backgrounds of some individuals are solely responsible for the barriers identified. As alluded to throughout our analysis, we must also examine the social context in which the experiences of the participants are embedded within. As such, a main question that has yet to be answered in this analysis, or in similar research, is: how are the experiences of our participants and others like them *organized to happen as they do*?[4]

In order to answer this question, we propose an in-depth institutional ethnographic examination of the work settings of participants, and how the work settings connect to and organize the experiences discussed above. As Campbell and Gregor (2002: 90) explain, "An institutional ethnography must therefore include research into those elements of social organization that connect the local setting and local experiences to sites outside the experiential setting" (see Smith (2005) for a discussion of institutional ethnography). While this question requires further investigation, it is the analytical claim here that the facilitators and barriers discussed above are inevitably connected to the work settings of participants. For example, how is the workplace setting organized or put together to reinforce the power imbalance where some individuals have power over others? As one participant argued, "the organization of management structure would put physicians above every other profession within the organization, and that needs to change before we can engage fully."

One of the benefits of using qualitative research strategies, and an institutional ethnographic lens specifically, to investigate interprofessional collaboration (and other elements of health care work and health care organization) is the ability to gain a better understanding of what people do and how what people do is organized to happen as it does (Smith, 2005). This is important because much of the actual work people do in health care is assumed or taken-for-granted and is not made institutionally visible (see Allen, 2014). In the context of IPC, for example, one participant explained that interprofessional collaboration is, "probably invisible. It's just understood that it happens, and accepted that it happens . . ." (Amna). Such taken-for-granted work is especially pervasive in the interprofessional education/collaboration literature; collaboration is assumed rather than making visible the multifarious ways collaboration is enacted and organized in practice. In other words, what collaboration actually looks like in practice and how collaborative work is socially organized vis-à-vis institutional policies and practices, regulatory frameworks, immigration and hiring practices, and accountability regimes, for example, is largely unexplored.

This aforementioned line of inquiry would complement pre and post-graduate interprofessional educational training in Qatar and elsewhere; investigating the social organization of healthcare in Qatar would allow us to better understand how the work settings in Qatar may, in "hidden" ways (Rankin and Campbell, 2006), work against both front-line workers and their patients. By understanding how the work of individuals in Qatar are organized by the institutional setting in which their work occurs, strategies for change can be identified and implemented in order to facilitate interprofessional education and collaboration in Qatar and elsewhere. Furthermore, an inquiry of this type would also aid in the development of a critical discourse of discussion in Qatar, which is necessary if interprofessional collaboration on the front lines of health care is to be realized.

Notes

1 This study was generously funded by the Qatar Academic Health System (EH/CSFAA/0068-2013). The authors report no declaration of interest. Additionally, we would like to acknowledge the support of Brad Johnson, Mohamed El Tawil, Kristen Mrus, Anne Topping, Alla El Awaisi, Marwan Hamad, Joanne Davies, Bernie Mulherin, The University of Calgary in Qatar, and Hamad Medical Corporation.

2 Sandelowski (2000, p. 335) aptly points out, "all inquiry entails description, and all description entail interpretation . . . there are no "facts" outside the particular context that gives those facts meaning. Descriptions always depend on the perceptions, inclinations, sensitivities, and sensibilities of the describer." Include this information in the text rather than in footnotes, per Harvard style standards

3 As an institutional ethnographer (Smith, 1987, 2005), the first author who conducted the interviews was drawn to explore some of the work strategies that participants deploy to activate working with other healthcare professionals. While this study is not an institutional ethnography, we classify "ethnographic interviewing" as utilizing some of the insights institutional ethnographers would use when interviewing individuals.

4 This is an institutional ethnographic question that, in the context of this research, seeks to better understand how what individuals do, or do not, is organized and coordinated by different text-mediated modes of social organization embedded in the work-setting of participants. We acknowledge Janet Rankin in the phrasing of this question (see also Campbell and Gregor, 2002; Smith, 2005). Include this information in the text or remove it, per Harvard style standards.

References

Al Jazeera. (2014) The plight of Qatar's migrant workers. Retrieved February 22, 2015, from: http://tinyurl.com/z8kfzuq

Allen, D. (2014) 'Re-conceptualising holism in the contemporary nursing mandate: from individual to organisational relationships', *Social Science & Medicine (1982)*, *119*, pp 131–138: http://doi.org/10.1016/j.socscimed.2014.08.036

Barrow, M., McKimm, J., Gasquoine, S., and Rowe, D. (2014) 'Collaborating in healthcare delivery: exploring conceptual differences at the "bedside"', *Journal of Interprofessional Care*, *29*(2), pp. 119–124: http://doi.org/10.3109/13561820.2014.955911

Campbell, M. L., and Gregor, F. (2002) *Mapping Social Relations: A Primer in Doing Institutional Ethnography*. Walnut Creek, CA: AltaMira Press.

Clark, P. G. (2011) 'Examining the interface between interprofessional practice and education: lessons learned from Norway for promoting teamwork', *Journal of Interprofessional Care*, *25*(1), pp. 26–32: http://doi.org/10.3109/13561820.2010.497751

Creswell, J. W. (2002) '*Research Design: Qualitative, Quantitative, and Mixed Methods Approaches* (2nd edition)', Thousand Oaks, Calif: SAGE Publications, Inc.

DeVault, M. L., & McCoy, L. (2006) 'Institutional ethnography: 'Using interviews to investigate ruling relations', in D. E. Smith (Ed.) *Institutional Ethnography as Practice*. Lanham, Md: Rowman & Littlefield Publishers.

Esterberg, K. G. (2002) *Qualitative Methods in Social Research* (1 edition). Boston: McGraw-Hill Humanities/Social Sciences/Languages.

Fellow, J. (2013) 'Qatar's delicate balancing act', Retrieved from: http://www.bbc.co.uk/news/world-middle-east-21029018

General Secretariat for Development Planning. (2008) *Qatar National Vision 2030*. Doha. Retrieved from: http://tinyurl.com/zqc83oc

Gill, P., Stewart, K., Treasure, E., and Chadwick, B. (2008) 'Method of data collection in qualitative research: Interviews and focus groups', *British Dental Journal*, 204(6), pp. 291-295.

Grant, H., and Falconer, R. (2013) 'Qatar's foreign domestic workers face abuse and exploitation, Amnesty warns', *The Guardian*. London. Retrieved from: http://tinyurl.com/jq78z8h

Institute for Health Metrics and Evaluation. (2013) 'GBD Country Profile: Qatar', Retrieved February 22, 2015, from: http://www.healthdata.org/results/country-profiles

Jones, G. M., & Delany, T. (2014) 'What Does Collaborative Practice Mean Within Mental Health Care?: A Qualitative Study Exploring Understandings and Proposing a Definition', *Journal of Research in Interprofessional Practice and Education*, *3*(3). Retrieved from: http://www.jripe.org/index.php/journal/article/viewArticle/154

Kritsotakis, G., and Gamarnikow, E. (2004) 'What is social capital and how does it relate to health?', *International Journal of Nursing Studies*, *41*(1), pp. 43–50: http://doi.org/10.1016/S0020-7489(03)00097-X

Moerer-Urdahl, T., and Creswell, J. W. (2004) 'Using Transcendental Phenomenology to Explore the "Ripple Effect" in a Leadership Mentoring Program', *International Journal of Qualitative Methods*, *3*(2), pp. 19–35.

Ng, M., Fleming, T., Robinson, M., Thomson, B., Graetz, N., Margono, C., … Gakidou, E. (2014) 'Global, regional, and national prevalence of overweight and obesity in children and adults during 1980–2013: a systematic analysis for the Global Burden of Disease Study 2013', *The Lancet*, *384*(9945), pp. 766–781: http://doi.org/10.1016/S0140-6736(14)60460-8 *not cited in text*

Qatar National Bank. (2014) *Qatar Economic Insight*. Doha. Retrieved February 13, 2016 from: http://tinyurl.com/zu2uj9r

Qatar Statistics Authority. (2015) 'QSA Population', Retrieved February 22, 2015, from: http://www.qsa.gov.qa/eng/populationstructure.htm

Rankin, J. M., and Campbell, M. L. (2006) *Managing to Nurse: Inside Canada's Health Care Reform*. Toronto: University of Toronto Press, Scholarly Publishing Division.

Reeves, S. (2011) 'Using the Sociological Imagination to Explore the Nature of Interprofessional Interactions and Relations', In S. Kitto, J. Chesters, J. Thistlewaite, and S. Reeves (Eds.) *Sociology of Interprofessional Health Care Practice: Critical Reflections and Concrete Solutions*. Nova Science Pub Inc.

Sandelowski, M. (2000) 'Whatever happened to qualitative description?', *Research in Nursing & Health*, *23*, pp. 334-340.

Smith, D. E. (1987). *The Everyday World As Problematic: A Feminist Sociology*. Boston: Northeastern.

Smith, D. E. (2005) *Institutional Ethnography: A Sociology for People*. Walnut Creek, CA: AltaMira Press.

Supreme Council of Health. (2011) 'National Health Strategy 2011-2016.

The economy: The haves and the have-nots. (2013)', *The Economist*. Retrieved from: http://tinyurl.com/h4222a2

Thistlewaite, J., Chesters, J., Reeves, S., and Kitto, S. (2011) 'A sociology of interprofessional Healthcare', In Health Care Issues, Costs and Access. In S. Kitto, J. Chesters, J. Thistlewaite, and S. Reeves (Eds.)

Sociology of Interprofessional Health Care Practice: Critical Reflections and Concrete Solutions. Nova Science Pub Inc.

World Health Organization. (2010) 'Framework for action on interprofessional education and collaborative practice', Retrieved December 14, 2013, from http://www.who.int/hrh/resources/framework_action/en/

World Health Organization. (2014), 'Ambient (outdoor) air pollution in cities database 2014', Retrieved February 22, 2015, from http://www.who.int/phe/health_topics/outdoorair/databases/cities/en/

12

Arab Expatriate Health Professionals

Heba Nassar

Introduction

The migration of highly skilled health professionals from developing to more developed countries has increased dramatically in response to a range of social, economic, and political factors. MENA is one of the most dynamic regions in terms of intra- as well as extra-regional migration. This paper seeks to analyze the nature, context, and patterns of health professional migration, particularly the migration of physicians and nurses in the region.

The paper is divided into four parts; part one serves as an introduction to the international and regional migration of health professionals. Part two examines selected cases including Egypt, Lebanon, Sudan, and the GCC. Part three examines the consequences of health professional migration with a special emphasis on the health and education systems in the Arab region. The paper concludes with policy recommendations.

Part One: Introduction: Migration of Health Professionals

1. Theoretical Background of Migration of Health Professionals

Migration can be defined as a permanent or semi-permanent change of residence, a movement in which people leave their country of origin. The focus in this chapter is on the migration of young and technically-skilled people, a phenomenon known as the brain drain. It is difficult to analyze the decisions that influence one to migrate. Most economic theories of migration rely on the so-called *push-pull model* for an explanation of the causes of migration. Lists of push factors that lead to migration include such elements as economic, social, and political hardships, while pull factors include the comparative advantages available in richer countries. Combinations of push and pull factors determine the size and direction of flows (Clemens, 2015).

Moreover, *neoclassical macro-economic* theory explains the development of labor migration within the process of economic development. Wage differentials, caused by differences in the ratio of labor to capital, encourage workers from low-wage countries to migrate to countries with higher wages. In addition to the benefits of expected higher wages, there are various costs, such as travel, wages forgone while looking for work, adapting to new surroundings, and the psychological costs of leaving friends and family. The larger the differences between the countries in terms of expected returns, the larger the size of the migration flow. Also the likelihood of finding work, as well as individual characteristics (education, experience, training, language skills), plays a vital role in determining migration rates (Straubhaar, 2000).

Though most of the studies refer to migration as being primarily motivated by economic factors, such as unemployment, low salaries, underemployment, and lack of opportunities, there are additional demographic factors such as age, gender, social class, level of education, family, social, and cultural factors that should also be considered. In this paper, we will exclude political factors, like political crises, discrimination, lack of political freedom, and many others as we will concentrate primarily on the economic factors.

2. Migration of Health Professionals: World View

Health workers, in general, are defined as people engaged in actions whose primary intent is to enhance health. These include people who provide health services – such as doctors, nurses, midwives, pharmacists, laboratory technicians – as well as management and support workers – such as hospital managers, and financial officers. The WHO estimates that there are about 60 million health workers worldwide where the majority (about two-thirds) provide health services; while the other one-third are involved in managerial and support work (World Health Organization, 2010).

The migration of health professionals has been on the rise; globalization has helped to drive this international migration. At the same time, demand for health workers has increased in high-income countries where not enough health workers are being trained locally and where the existing workforce is aging.

In general, many factors contribute to the migration of health workers. Pull factors include: better working conditions, better career opportunities, greater job satisfaction, higher quality of management, and governance. On the other hand, there are push factors in the receiving countries, which contribute to the migration of the health professionals, such as poor working conditions, low wages, high unemployment rates, political instability, and conflicts.

3. Regional Arab Migration: Historical Trends

Past research has classified Arab migration according to its two distinct migratory patterns: intra-regional migration and extra-regional migration. The latter is considered the South-North trend in migration from Arab countries to OECD countries.

Intra-regional Arab migration, or the so called South-South Context of Migration, can be classified into three types: the first type is the one that takes place between some Arab countries and other countries that are not located in the Gulf area such as Libya. The second type is a phenomenon where the same sending country is also the hosting country of migrant labor, as with the case in Jordan (replacement migration). The third type is the traditional type of migration, which takes place between the non-Gulf labor exporting countries and the Gulf labor importing countries (Nassar et al., 2003).

Moreover, MENA is a major exporter of educated professionals to the outside world. The United Nations Economic and Social Commission for Western Asia (ESCWA) has estimated that the majority of international migrants aged 15 years and above reside in OECD countries. The results indicate that of the ESCWA member countries, Lebanon, Iraq, Egypt, and to a lesser extent, Syria, have the largest number of educated migrants residing in OECD countries (Economic & Social Commission for Western Asia, 2009).

These results are affirmed looking at OECD data from 2000. Egypt, Kuwait, Qatar, Palestine, Bahrain, Jordan, and Sudan represent the countries with the highest number of educated graduates living in the OECD – two-fifths to nearly half of their migrant populations in 2000 held college degrees. They are followed by Oman, Saudi Arabia, Syria, Lebanon, and Djibouti, where 30% to 40% of their migrant population in the OECD hold certificates. The Maghreb countries (Algeria, Tunisia, and Morocco) come in last place since only 15% to 16% are credentialed (Database on Immigrants in OECD Countries (DIOC), 2010).

4. Health Professionals Migration in the Arab World

The migration of health professionals in the Arab region is on the rise. The direction of this migration is from countries with large populations in labor exporting countries such as Egypt, Sudan, Algeria, and Syria to countries with smaller populations that have historically imported labor, namely the GCC. Accelerated economic growth in the GCC is increasing the demand for educated professionals in general. Human capital production in the Gulf is insufficient to meet these growing demands. Hence, these countries have grown dependent on importing foreign-trained health professionals from Lebanon, Jordan and Egypt, or countries further away, such as India, Pakistan, and the Philippines (El-Jardali et al., 2008).

Common languages and cultural ties also contribute significantly to the movement of Arab health professionals. An Egyptian private recruitment agency that specializes in health workers reported that 90 percent of its work volume is in Saudi Arabia, the rest are in Bahrain, Kuwait, Qatar, and in Egypt itself. Similarly, the vast majority of the 60% of Sudanese doctors and 25% of Sudanese pharmacists working abroad do so in Saudi Arabia and other Gulf states (The Global Policy Advisory Council, 2010).

Part Two: Health Professionals Migration in the Arab World: Case Studies

1. Receiving Countries: GCC Region

A. Overview of Migration in the GCC Region

The so-called "South-South" model of migration is most common in the GCC. Although GCC countries have limited inflows of migrants from the OECD, it is mostly Asian and Arab labor migration that represents the majority of migratory flows to the GCC (Table 2-a).

The number of migrant workers in the GCC has increased from 9 million in 1990 to about 24.2 million in 2010-14, nearly half of the population in the GCC region (Table 2-b). The GCC region hosts about 9.7% of total migration in the Arab region and about 9.7% of total international migration all over the world, making the GCC region one of the most attractive regions for migration internationally (Abu ElNasr, 2014). Between 2010-2014, the percentage of total non-nationals to the total population in GCC countries ranged from 88.5% in U.A.E. to about 44% Oman and 33% in Saudi Arabia (Table 1 in Appendix).

Foreigners are over-represented in the private sector while nationals primarily hold jobs in the much more regulated public sectors (Suter, 2005). According to the most recent available data, the percentage of non-national workers averages 88.2% in private sectors leaving less than 10% in the public sectors over 2010-2014 (Gulf Labour Markets and Migration Statistical Database, 2015).

In all GCC States, Arabs now constitute only a minority of the expatriate labor force, constituting about 23% of total expatriates in the GCC region. The highest share, at 30%, is in Saudi Arabia. On the other hand, Asian migrants dominate the foreign labor market in all the GCC countries. They make up about 80% of the labor force in all GCC countries except for Oman where their percentage is higher (93.4%) and about 70% in Saudi Arabia (Table 2 in Appendix).

B. Health Professional Migration in the GCC Region

Based on El-Jardali et al. (2008) analysis the majority of health workers, mainly physicians and nurses in the GCC countries are expatriates from different countries, mostly from areas such as Southeast Asia, MENA, North America, and the U.K. In the U.A.E, it was estimated that about 80% of health workers are expatriates and less than 20% are nationals, whereas 63% of the nurses working in the Ministry of Health came from East Asia (mainly India, Pakistan, and the Philippines), while 28% came from Arab countries (Palestine, Jordan, Oman, Syria, Egypt, Sudan and Somalia). The percentage of nationals varies according to the sector, as it was relatively high (24%) in the ministry of health facilities, a percentage much higher than the percentage of nationals working across the U.A.E (7%). This may be due to the fact that the government is trying to employ as many nationals as possible. Finally, according to current estimates, there are three times as many expatriate health professionals as national health professionals in the GCC (Table 3 in Appendix).

Two explanations are typically cited for this imbalance. First, inadequate incentives are often pointed to as a cause of disinterest. For many GCC nationals, the many years of study and preparation required for entry into medical professions outweigh the benefits of working in the health sector. Second, social and cultural reasons may dissuade some GCC nationals from healthcare professions such as nursing that require long work hours and night shifts (El-Jardali et al., 2008).

C. Socioeconomic Implications of Labor Mobility in the GCC Region

The GCCs went through a period of rapid population growth between the 1940s and the 1980s. Since then, fertility rates have steadily declined, but the wide-based age pyramid created in that population boom produced a 'population momentum' that will yield high growth over the coming decades. Currently, forty-two percent of the population in the Gulf area is under the age of 15. The data indicate a fast growing workforce in which the number of entrants grossly outweighs the number of retirees. Social reforms involving the educational advancement of women, who have traditionally been excluded from the labor market, have further increased the workforce growth rate. The mismatch between the growing national workforce and the number of opportunities for available employment manifests itself in steadily increasing unemployment rates. Increasing pressure on the labor market poses economic, social, and political challenges for GCC governments.

As unemployment among nationals begins to grow – a phenomenon unheard of in the past – the GCC governments have decided to embark on the formulation of labor market strategies to improve this situation, to create sufficient employment opportunities for nationals, and to limit the dependence on expatriate labor (the so-called localization, nationalization or indigenization of labor, which, depending on the country, is referred to such as Saudization, Omanization, Emiratization, etc.). A number of measures have been proposed to achieve these objectives: some professions have been reserved as 'for nationals only.' Employment quotas for nationals and expatriates have been introduced in certain professions, and wage subsidies and state retirement plans have been established for nationals in the private sector in order to parallel fees and charges on foreign labor to make migrant laborers less competitive. Private companies that meet quota requirements have been rewarded. Moreover, large efforts have been made to improve the education and training of nationals.

As a result of these policies, there has been noticeable progress made toward campaigns to nationalize jobs in both private and public sectors. Available evidence from Oman, Saudi Arabia, and Kuwait suggests that the Omanization, Saudization, and Kuwaitization drives have yielded positive results. The private sector employed 50,923 more Omanis during the period between 1996-2000 as compared to only 10,846, indicating a strong reversal of past trends. Nevertheless, while the public sector has become successfully nationalized, all these measures have so far brought limited results in the private sector. Due to long work hours, low salaries, and overly competitive work environments the private sector remains unattractive for nationals (Nasra, 2005).

Within the health care sector, many recommendations, such as better salaries, professional training residencies, and more flexible careers, have been implemented to encourage GCC nationals to work in medical professions. Qatar has entered partnerships with European and North American medical schools to enhance its medical training. Similarly, the Bahraini government collaborated with the Royal College of Surgeons to open a postgraduate medical training institute in Bahrain. Due to these initiatives, many nationals may reconsider the attractiveness of health care professions. However, in spite of these policies, the number of new medical graduates will not keep up with the GCC's population growth (Mourshed et al., 2006).

2. Health Professionals Migration in the Arab World: Sending Countries

This chapter assesses migratory flows from the following three sending countries: Egypt, Lebanon, and Sudan.

A. Overview of Migration Patterns from Egypt

Egypt's principal migration legislation, law number 111 of 1983, makes a clear distinction between permanent and temporary migration. In practice, the distinction is simply a geographical one, with all migrants to Arab states defined as temporary, even though some have been there many years. This also relates to the labor migration regime in the GCC countries, where they do not allow provisions for permanent settlements and migrants are not able to gain citizenship no matter how long they may have stayed in the country. On the other hand, all migrants to Europe, North America, or Australia are defined as permanent, including those who have arrived very recently.

According to the latest data available from the Egyptian Ministry of Foreign Affairs, the total number of Egyptian migrants abroad reached 6.5 million in 2009. Arab countries host the most, about 4.8 million Egyptians (74%) followed by North America (12%) and the European countries (12%) where each hosts about 800 thousand Egyptians (Table 4-a in Appendix). Egyptian labor migration to Arab countries (i.e. temporary migration) consists of two main groups: The first group includes non GCC countries, primarily Libya and Jordan. The second group includes the GCC countries.

Table 4-b refers to the distribution of the number of Egyptians, who have obtained work permits grouped according to country of residence. Between 2001-2013, Arab countries represented the main destination for temporary Egyptian Migration, averaging about 96% for this period.

Centrak Agency for Public Mobilization and Statistics, 2007 estimates that, the GCC countries hosted about 85% of total Egyptians working in MENA. Saudi Arabia, Kuwait, and the U.A.E. were the main destinations for Egyptian workers. Nearly half of Egyptian migrant laborers to Arab countries headed to Saudi Arabia (47.5%), followed by Kuwait (20%), and the U.A.E (11.4%). On the other hand, Jordan received about one-sixth of the Egyptian migrants to Arab countries in 2007 (15.4%) (CAPMAS Statistical Database, 2007).

B. Health Professional Migration in Egypt

Common destination countries for Egyptian health professionals are mainly the Gulf states as well as Western countries like the US, Canada, Australia, New Zealand and the EU. Migrants to the Gulf States could be attracted by better financial returns. However, Egyptian health professionals migrate to the West mainly to pursue further education, knowledge, and expertise.

The Gulf States are the most likely destination sought after by Egyptian health professionals as their Egyptian degrees are accredited in the Gulf. They earn better wages and accumulate more wealth, after which they either wish to return to Egypt, having secured savings, or use this money to finance their migration to Western countries (Yasmine et al., 2012). As a third step, they could return back to the Gulf to be hired as western-trained health professionals.

Fouad et al. (2014), used an online questionnaire to survey 940 Egyptian undergraduates enrolled at the Faculty of Medicine, Ain Shams University to explore their intentions of mobility following graduation and the factors impacting such intentions. The survey results reflected that 85.7% of the responding students intended to leave the country for training following graduation, mostly seeking better research opportunities and working conditions. Among those who intended to train abroad, the three most mentioned destination countries were: the USA (29.8%), Germany (26.8%), and the UK (20%), followed by Canada (8.1%) and Australia (5.7%). The majority of those who intended to train abroad stated their plan was to eventually return to their country (although after different intervals of stay): 12.8% planned to return directly after training, 26.9% planned to stay abroad for less than five years before returning, 26.4% planned to stay abroad for 5–10 years before returning, and 15.6% planned to stay abroad for more than 10 years before returning. Only 18.2% planned to stay abroad indefinitely, 66% of whom were males (Youssef et al., 2014).

Although there is no available national data on the migration of the health professionals in Egypt; the available statistics could, however, indicate some important results (Table 7-a-b in Appendix).

- Based on the occupational structure of Egypt's permanent migrants, i.e. those who headed to developed countries in 2013, it is evident that more than two-fifths of the permanent Egyptian migrants in 2013 were highly skilled specialists (41%).

- According to the data on permanent migration by specialization in 2013, it is evident that graduates with commercial (29%) and engineering (22.3%) backgrounds constitute the largest group of Egyptians who moved permanently to another country (Table 7-b). They are followed by health professionals (21.9%) such as pharmacists (14%), physicians (5.7%), and dentists (1%); however, there is no clear statistical estimate of the migration of the nursing staff.

The low number of health professional migrants to Western destinations is due to the fact that obtaining a work permit to practice clinical medicine is very difficult there. Table 8 reflects the estimates of Egyptian doctors and nurses working abroad. The evidence illustrates that the U.S.A., followed by the U.K., represents the main destination country for Egyptian physicians; it also shows that the movement of Egyptian nurses to the developed Western world is limited as the study assumed that the number of Egyptian nurses in these countries did not exceed 1,000 nurses (Clemens et al., 2008).

A point worth mentioning is that the mobility of Egyptian health professionals towards Western destinations has been more limited since the mid-1990s, particularly to European countries, after the establishment of the European Union. This change is due to increased mobility restrictions including language requirements, the need for scholarships, and the ability to cover one's living expenses.

As for the migration of Egyptian Nurses, the Gulf States represent the sole destination for Egyptian nurses wishing to migrate. Nursing education in Egypt is considerably underdeveloped. The majority of Egyptian nurses are diploma holders of secondary level, technical, nursing schools and, to a lesser extent, the High Institute of Nursing in Alexandria. The fact that most nurses do not have higher levels of training makes it difficult for them to pursue any work or advanced training in EU countries while the top nursing students are employed by the university as faculty members (Moataz et al., 2012).

3. Health Professionals Migration in the Arab World: Sending Countries: Lebanon

A. Overview of Lebanese Migration

Lebanon has the longest history of migration among its Arab counterparts. This Lebanese migration started in the second half of the 19th century; the first migration movements were in response to a number of factors, in particular, the religious communal conflicts and the economic crisis in Lebanon. Migration was concentrated in Latin America and, to a lesser extent, Europe and the U.S. Later, during most of the twentieth century, many Lebanese migrants headed to the U.S., Canada, Australia, and France permanently, as well as to Western Africa temporarily. Since the 1960s, there has been increased Lebanese migration to the Gulf states. The Civil War of 1975-1989 led to an increase in immigration to all the countries and regions, which included migrants with varying socio-economic profiles.

Lebanon's labor market has constantly been characterized by high rates of brain drain and unemployment, especially during times of political instability and civil strife. In particular, the migration of the young, educated workforce, especially to the Gulf, has always been a problem for Lebanon, as most Lebanese view the Lebanese labor market as lacking economic opportunities. Data on the number of Lebanese migrants and their main characteristics are fragmented. However, estimates show that 46.2 % of Lebanese households have at least one close family member who migrated from 1975 to 2001. Reportedly, between 600,000 and 900,000 people migrated from Lebanon between 1975 and 2001; the average migrant is usually single and is twice as likely to be male rather than female. North America was the predominant destination for the Lebanese migrants (33%), followed by Europe (25%), Arab countries (20%), and Australia (13%). Two-thirds of Lebanese migrants have permanently settled in their host countries and/or hold dual citizenship. Half of the respondents cited employment as the reason behind their migration: 17% left because of the general situation in the country, 20 % for family reunification, 9% to pursue education abroad, and 5% because of the war. A third of the Lebanese who emigrated during the war returned at a later stage during the war (1975-1990), 26% returned between 1991 and 1995, and 27 % returned between 1995 and 2000 (Tabar et al., 2013).

According to a survey carried out by St. Joseph University (USJ) in 2007, the number of Lebanese emigrants between 1992 and 2007 is estimated at 466,019. At that time, almost half of Lebanese households (46.0%) had at least one emigrant among its close kin. Arab countries represented the main destination countries for Lebanese migrants at 35% of the total Lebanese migrant population between 1992-2007, or approximately 162,663 migrants (Di Bartolomeo et al., 2010).

B. Health Professionals Migration

Lebanon has a physician migration factor of 19.3%, the 7th highest in the world as well as the highest in the Middle East and North Africa. Since 1978, the number of Lebanese medical graduates practicing in the United States (US) has been consistently rising (Akl et al., 2008).

According to the data available, the physician density in Lebanon is almost thrice the nurse density, indicating a major professional mal-distribution. Curiously, the growth rate in the supply of physicians is 9%, which exceeds the average rate of population growth by 1.9% per year (El-Jardali et al., 2008-b).

An estimate of the migration intentions of Lebanese medical students found that 96% of respondents intended to train abroad either for a specialty (78%) or a subspecialty (18%) (Akl et al., 2008). The top four destination countries were the U.S. (74.1%), France (12.1%), the United Kingdom (7.6%), and Canada (4.2%). While (25.1%) of those who intended to train abroad wanted to return to Lebanon directly after finishing training, 63.8% intended to return after working abroad for a varying number of years. Interestingly, (10.6%) intended to never return to Lebanon (Akl et al., 2008).

Lebanon suffers a severe nursing shortage. The nurse density in Lebanon is one of the lowest in the world; the World Health Organization (2006) reported that the nurse density in Lebanon is 1.17 per 1000 population compared to a regional average of 2.20 per 1000. The shortage results from an unattractive professional status, high turnover, poor job satisfaction, poor work environment, and excessive emigration to the Gulf region. In addition, about 67.5% of a total of 1,793 nurses who were included in the survey reported an intent to leave within the next three years. Financial reasons were the main justification for nurses leaving for the Gulf, whereas continuing education was the main reason for leaving for nurses in North America and Europe (El-Jardali et al., 2008-b).

4. Health Professionals Migration in the Arab World: Sending Countries: Sudan

In assessing Sudanese migrants, Di Bartolomeo et al. (2012) relied on statistics from destination countries for which the population distribution by nationality is available. According to this data, in 2010, 907,544 Sudanese migrants resided abroad. The Gulf region hosts the majority of the Sudanese migrants (71.9%). Saudi Arabia represented the major destination country for Sudanese migrants (551,768 migrants). Sudanese migrants in the developed world mainly reside in the U.S., the U.K., and Canada (38,166; 20,000 and 12,590 migrants consecutively) (Assal, 2010).

There is a lack of detailed data about the characteristics of the skilled Sudanese migrants working abroad. However, the 2008 data shows that the great majority of Sudanese expatriates work in skilled and semi-skilled professions. The total number for this category of skilled and semi-skilled is 451,443. The data also show that there is a significant number of expatriates, who hold high professional positions such as engineers, medical professions, university professors, and legal professions (Assal, 2010).

In Sudan, the migration of health professionals has been increasing since the 1960s. Current estimates show that over 60% of Sudanese physicians practice outside the country (ElSheikh et al., 2006). Clemens, Michael & Pettersson (2008) estimated that the number of Sudanese health professionals abroad was relatively high ranging from about 13% of the total domestic staff; however, the movement of Sudanese nurses to the developed Western world is limited (Clemens et al., 2008).

Push factors for Sudanese health workers are similar to the factors for their counterparts in other sending countries – looking for better wages, improved working conditions, escaping political conflict, civil wars, better education, knowledge and training are some examples (ElSheikh et al., 2006).

During the 1980s and early 1990s, the Sudanese Ministry of Health faced a shortage in medical personnel so dire that the Ministry banned the migration of physicians. Yet, this restriction did not curb migration. Physicians, who were determined to migrate simply changed the professions listed in their passports (ElSheikh et al., 2006). The brain drain among pharmacists is considerable, involving 25 percent of the country's total population of pharmacists. The migration of nurses is not common; one reason is that highly-skilled nurses are few in numbers. Female nurses face socio-cultural and legal hurdles that discourage them and in some cases this limits the migration of unaccompanied females (Assal, 2010).

Part Three: Impact of Migration

Migration of health workers has negative and positive effects, as follows:

1. Negative Consequences

A. Shortage of Human Resources for Health

Like other developing countries, the migration of the health professionals from the Arab sending countries constitutes a major loss of a significant number of nurses, physicians, and other healthcare professionals. Healthcare systems in most of these countries already face severe problems such as inadequate funding, inferior technology, epidemics, a lack of infrastructure, insufficient training capacity, and a long-standing shortage of healthcare professionals. The further loss of nurses and physicians to developed countries damages the already poor health care systems in these countries. The impact is particularly significant when such personnel cannot be replaced due to shortages. While recruiting foreign workers may help host countries in overcoming staff and skill shortages, it deprives source countries from essential knowledge, skills, and expertise (Stewart et al., 2007).

In 2011, The World Bank estimated that the region has lost 27,265 physicians to migration, which is nearly 7.8% of physicians trained in the region (World Bank 2011). For example, it is estimated that 36% of outreach health facilities in Sudan are not functioning and there is only 55% coverage of basic health services in the country. Sudan loses approximately US $24 million annually in money spent to train physicians who then migrate· (ElSheikh Badr and Abd Allah, 2006).

Despite the excess of physicians in Lebanon (a density of 325 doctors per 100,000), the loss of medical expertise is still potentially damaging. This migration is associated with the loss of intellectual capital and of educational investment with uncertain returns. Secondly, there is indirect evidence that those who migrate might be the best (Akl et al., 2008). This possible selection process could negatively affect the quality of health care services in Lebanon. Moreover; the oversaturation of the national medical labor market could be discouraging migrant Lebanese physicians in other parts of the world to consider resettling in Lebanon, which would likely exacerbate Lebanon's brain drain problem (Akl et al., 2008).

What complicates the problem further is that many Arab countries already face a severe challenge in terms of the geographical distribution of their medical labor force. Many Arab countries suffer from the spatial mal-distribution of health workforce across different regions, urban and rural, across different health institutions such as hospitals and primary health care facilities, as well as across public and private sectors. This phenomenon is further exacerbated by the lack of incentives to encourage physicians to practice in these areas (El-Jardali, 2008-b). For example, in Egpyt, health professionals are concentrated in urban centers like Cairo, Alexandria, and Upper Egypt, but there is a shortage of health professionals in Lower Egypt. Moreover, health workers prefer working in private healthcare centers and urban areas, which is attributed to better salaries (Egypt Demographic & Health Survey, 2008).

B. Brain Drain

The migration of health professionals and qualified workers contributes to the brain drain in MENA. It effectively deprives them of trained personnel, whose education they have invested in, thereby exacerbating the existing dearth of highly-skilled human resources. The source countries incur the loss of the cost they bear in providing their health professionals with training. The training of healthcare professionals in most developing countries, including many of the countries sending Arab countries, is either entirely sponsored or heavily subsidized by the government. This substantial investment in training is lost when a nurse or a physician permanently migrates to a developed country. For example, the United Nations estimates that each migrating African healthcare professional represents a loss to the source country of $184,000 (World Bank, 2010).

C. Impact on the Education System in the Arab Countries

Undergraduate and postgraduate medical education suffers as a result of migration, affecting the next generation of health professionals in some source countries. Specialized health professionals who migrate are often among the few active or published researchers in low-income source countries. The migration of such individuals stifles innovation and invention in dealing with persistent local public health problems such as hepatitis B & C, HIV/AIDS, tuberculosis, and malaria (Packer et al., 2009).

On the other hand, education spending is high in some countries with excessive levels of brain drain: Algeria, Morocco, and Tunisia all spend above the regional average. In Morocco and Tunisia, tertiary level spending is especially high in absolute and relative terms. These investments are thus lost as students go abroad to work (World Bank, 2010). Moreover, skilled migrants do not always end up in skilled jobs abroad. This hinders the potential that skilled migrants can acquire host country knowledge, and thus can help transfer such knowledge back home and get a higher private return on their investment in education than they would have had in their home countries. However, some highly educated migrants do end up in low-skilled occupations. Moreover, as the education systems in most Arab countries already face many challenges that affect the development process in these countries, accordingly, the loss of these highly skilled migrants could further aggravate these challenges (World Bank, 2010).

2. Positive Consequences

On the other hand, migration of health professionals could generate positive results mainly manifested in the possibility of knowledge and skills transferred by the migrants as well as the financial returns of migration in the form of remittances the migrants send back to their home countries.

A. Knowledge & Skills Transfer

In theory, the temporary migration of healthcare workers can benefit their countries of origin. For example, when nurses and physicians leave their home countries and move to more developed countries, they could gain experience and training in a more advanced setting which would benefit their countries upon their return.

For instance, in the United Kingdom, Egyptian medical professionals have established a professional non-profit association named the Egyptian Medical Society. The society includes 120 members, many of whom are doctors residing in different parts of the United Kingdom, a country that hosts a large number of Egyptian skilled professionals. One of the society's goals is to contribute to the training of Egyptian medical personnel in Egypt and in the UK. There is also the Arab-German Medical Alumni Network (AGMAN), which is also a medical-based network that is comprised of almost 250 alumni based in Egypt who have studied in Germany either through a scholarship or on their own. Their activities range from forging projects with the host institutions in their migrant communities, holding lectures and cultural events at their home institutions and providing pre-departure training and orientations for potential emigrants (Moataz et al., 2012).

However, the assumption of possible knowledge and skills transferred due to migration of its highly skilled professional could face some limitations in the Arab region context. On one side, there is little evidence that more than a small percentage of migrants actually return. Even in cases where healthcare workers do return, healthcare systems in most of the sending Arab countries may not be able to take advantage of the skills and expertise acquired abroad. The technologies available in many of these countries' health systems would be much less sophisticated than those in developed countries, thus reducing the potential of benefiting from the skills and experience obtained overseas (Stewart et al., 2007).

B. Remittances: Sending and Receiving Countries

Receiving Countries

There has been much debate on the developmental role of remittances. Remittances can contribute to reducing inequalities that result from globalization; migration can improve the welfare of the country of origin as migrants may accumulate savings overseas that, given the low wages and capital market distortions, might not have been possible without migrating. Moreover, attractive investment opportunities in the country of origin can capture remittances for the stimulation of production and employment (Nassar, 2006).

Though the Arab region's share of global trade is below 5%, it is a major recipient of remittances; the share of the Arab region's total remittances to the developing countries has been rising, registering $47 billion in 2012 compared to $US34 billion in 2012. The Arab region's share of total remittances in the developing world averaged 13% over 2009-2012. This share is estimated by the World Bank to increase to about 18% by 2015 (Table 8 in Appendix).

The Arab region is unusual in that some of the larger remittance-recipient countries are close neighbors to some of the larger remittance-source countries. Remittances from GCC countries represent an important source of remittances within the region. In Morocco, remittances come mainly from European countries (World Bank, 2008). Jordan is a special case because it is an important receiver and sender of remittances in the Arab region. In 2014, Egypt maintained its first place with remittances advancing to $18 billion. Lebanon and Morocco stand as the second- largest recipients, with remittances averaging about $7.7 billion and $6.9 billion respectively. As a share of GDP, however, Lebanon and Jordan are most reliant on this income source (from the GCC countries registering 17 % and 10.8% of output in 2014 consecutively (Fig. 2 in Appendix)).

The Developmental Role of Remittances on Education and Health

The first channel of impact is a direct channel where remittances sent back from migrants are believed to alleviate household liquidity constraints and hence encourage receiving households to invest in their dependents' education (Chaaban et al., 2012).

A key finding in the literature is the positive impact of migration and remittances on education outcomes, particularly among younger children, girls, and children whose parents' education is low. This positive outcome of migration is because the marginal value of the remittance income would be higher for such groups and hence its effect on budget constraints and education spending would be stronger. No data is available on the amount of monetary or intangible remittances Arab migrant health professionals send home. However, this paper tries to use the available empirical studies on the impact of remittances on the health and education in remittances recipient countries. To estimate the impact of migration and remittances on schooling in Egypt, El Badawy et al. (2009) depended on data from the Egypt Labor Market Panel Survey (ELMPS 06)[1]. The main results can be summarized as follows:

With respect to the effects on schooling attendance, while there was not much of an effect of migration indicators on young boys 6-11 and 12-14; there was only a mild positive effect on remittance recipients on school attendance for boys aged 15-17. On the other hand, results showed that remittances have a strong positive effect on attendance for university-aged boys. The unconditional probability of attending school for the average university-aged boy is 30%. This probability increases by 20 percentage suggesting that when the household receives remittances there are positive outcomes.

With respect to the effect of migration and remittances on schooling of girls, there has been little evidence that migration and remittance recipients increase the likelihood of attending school. For young girls 6-11, the positive migration effect can be interpreted as a speeding-up of school attendance since they are less likely to be late school-entrants. There was also some positive effect of migration and receiving remittances on school attendance for older girls. Like in the case of boys, remittances may increase the probability of school attendance for university-aged girls. However, the result is not as robust as it is for boys.

Chaaban, Jad & Mansour (2012) *assessed the impact of remittances on education in Jordan, Syria, and Lebanon. The results of the study indicated that:*

There is a positive impact of remittances recipients on education attendance for individuals aged 15-17 only for Syria. The results also showed that remittances have a significant impact on males aged 18-24 but not on females of the same age bracket, except for Lebanon where the reverse is found. For females in Jordan and Syria, two factors play a key role in preventing education attendance: age and marriage.

In general, migrant remittances encourage both females and males aged 18-24 to reach higher levels in their education. The effect appears to be larger in magnitude for males in cases from Jordan and Syria while for Lebanon the reverse trend is observed.

Remittances have a positive impact at the individual and household and for governments who viewed remittances as an easy source of hard currency. However, the Arab countries have not been successful in establishing well-organized institutional frameworks to mobilize remittances effectively for saving and investment purposes, in contrast to the policy and institutional frameworks of countries such as South Korea, the Philippines, and Thailand.

Sending Countries

The Gulf region is the third largest labor receiving region in the world, behind only North America and Europe, which have historically been destinations for migration. This is reflected in the significant share the GCC contributes to the overall remittances in the region. Remittances typically fund their homecoming. Put together, the GCC countries' official remittance crossed the $75 billion mark in 2012, which is 50 percent larger than that of the US, the top remitting country. The share of remittance from the GCC countries is around 20 percent or one-fifth of the total documented remittance in the world for 2010, 2011, and 2012. Furthermore, the relative size of these changes from the GCC countries is the largest among remitting countries (Naufal et al., 2014).

The large size of remittances from the GCC is not only linked to the significant size of the migration in the region but is also related to the general characteristics of the legal migration framework in the GCC. In general, migrants in the GCC countries have limited investment opportunities; for instance, they cannot own property in most countries, they face strict restrictions for bringing family members, and they usually do not have any opportunities to become citizens. These constraints make all migrants in the region temporary workers, contract workers and, thus, expatriates who, regardless of their nationality or skill level, share the objective of preparing for their eventual return home. These limitations create significant incentives to send money back home rather than use it in the local economy (Naufal et al., 2014).

Part Four: Policy Recommendations

1. Creating Human Resources for Health Strategies

As indicated in this paper, most Arab sending countries face several problems in their health systems, many of which come from the management of the available human resources for health. Apart from having an excess or shortage in health workers it is clear that most Arab countries still lack a clear strategy to understand health workforce challenges. This includes:

1. Conducting a needs assessment to establish a system that targets population needs and anticipates future needs of medical professionals. This involves needs-based HRH planning to address shortages in numbers, specialties, and spatial and sectoral mal-distributions.
2. Creating a national database for HRH that can be updated regularly. Such a database could provide accurate estimates of the current workforce and future needs.
3. Revising educational curricula to reflect international standards as well as the health needs of the population.
4. Improving and enhancing training programs available for health professionals through the provision of updated continuing education programs.
5. Creating financial and non-financial incentives; financial incentives could include better wages and remuneration systems whereas non-financial incentives could include improvement in practice environments and instituting a rewards and recognition system.
6. Managing shortages in nursing: shortages in nurses in many Arab countries are not only related to the migration of nurses but to the wider framework of managing human resources for health care in the Arab region. Accordingly, there is a need for developing a national strategy for nursing in the Arab countries including:

- Recognize nursing as an area of national need.
- Develop a collaborative strategy with input from educators, clinicians, administrators, and policy makers to meet the nursing workforce challenges.
- Organize national campaigns that inform the public of the value of the profession.
- Revise the nursing law that regulates the profession.
- Improve working conditions of nurses such as working hours, remuneration (salaries/benefits), and training opportunities.

2. Addressing the Push Factors for the Health Worker Migration

1. Develop a system for managing migration: there is an urgent need to develop a system to manage the excessive migration of the health workforce through establishing Human Resources for health retention strategies at a national level that can help retain the current workforce especially in rural and remote areas (WHO 2010).
2. Addressing the "push" factors that motivate healthcare workers to emigrate, which include improving wages, compensation, and working conditions. In addition, retention strategies could be adopted which include non-wage instruments like improved working and living conditions, which could be more effective in controlling migration in these countries. Other than financial incentives, the provision for professional development has worked effectively in Ghana and South Africa. Health professionals and students from local areas were awarded scholarships to undergo training in high-income countries if they agreed to return and work in rural and underserved areas of the country (Manisha et al., 2012).
3. Brain circulation: an innovative approach taken by China since 2001 is 'brain circulation'. The Chinese government is encouraging their 'lost talent' to return for short-term assignments or hold concurrent positions in China and abroad to aid research and development in the country. A similar trend is occurring among Indian emigrants (Manisha et al., 2012).

3. On the Destination/Receiving Countries Side

Responsible recruitment policies could be established by destination/receiving countries and fair treatment of migrant health workers.

4. On the International Level

Encouraging sending and receiving countries to adopt and implement the WHO Global Code of Practice on the International Recruitment of Health Personnel (WHO, 2010).

In 2004, the World Health Assembly requested the WHO to develop a code of practice for the international recruitment of health personnel. In response, the WHO initiated a global consultation process to produce a draft code. The code was adopted by the World Health Assembly in May 2010.

The Code of Practice is voluntary, global in scope, and applies to all health workers and stakeholders. It sets out principles and encourages the setting of voluntary standards. The equitable balance of the interests of health workers, source countries, and destination countries is promoted, with a particular emphasis on addressing the negative effects of health worker migration on countries experiencing a health workforce crisis. Key components of the Code include:

- Greater commitment to assist countries facing critical health worker shortages with their efforts to improve and support their health workforce.
- Joint investment in research and information systems to monitor the international migration of health workers in order to develop evidence-based policies.
- Member countries should meet their health personnel needs with their own human resources as much as possible thus take measures to educate, retain, and sustain a healthy workforce.
- Ensure that recruiters and employers observe fair and just recruitment and contractual practices in the employment of migrant health personnel and that migrant health personnel are not subject to illegal or fraudulent conduct. Migrant health personnel should be hired, promoted, and remunerated based on their qualifications, such as level of experience and professional degrees. This responsibility will be on the countries implementing equal of treatment the domestically trained health workforce.
- Migrant workers' rights are protected and equal to domestically-trained health workers (WHO, 2010).

5. Bilateral and Regional Cooperation

1. Establishing a body that will be responsible for designing, developing, implementing, and following up pan-Arab legal instruments for employment across Arab countries. This body will ensure the application of standard contracts with detailed and clear specifications of the duties and rights of both employees and employers. In this regard, an Arab Labor Court comprised of judges from both labor-sending and receiving Arab countries would be responsible for fast and final judgments in relation to disagreements and disputes that may arise between employers and employees (Haque et al., 2013).
2. Regional dialogue and cooperation between the countries of origin and destination countries should be the aim for sustainable co-operation. Dialogue should place migration into a broader framework aimed at creating a suitable environment for regional cooperation. Dialogue should assure the coordination within governments and consultation with all social partners on labor, development, and migration policies.
3. Enhance bilateral and regional dialogue to better manage the migration of health professionals in a way that fulfills the interests of the source countries in meeting the shortages in their national health workforce as well as enabling the sending countries to better manage and coordinate their own national workforce.

Some bilateral agreements were concluded between European Union member states and Arab-Mediterranean origin countries. France has concluded bilateral agreements with Tunisia and Morocco, among others, on migration and development. These bilateral agreements address the temporary migration of nationals of both countries and their visa requirements, coordination on combating irregular migration, and formation of partnerships for "co-development." Bilateral agreements concluded between France and Morocco also address the temporary migration of young professionals for a period of 3-12 months to improve their work experience and linguistic abilities. Spain has also concluded a number of bilateral agreements with Arab countries. Egypt and Tunisia also have concluded bilateral agreements on temporary labor migration with Italy (ILO, 2009).

6. Brain Gain

Maximizing the Potential Benefits of the Migration of Highly Skilled Professionals on the Health and Education Systems in the Arab Sending Countries through fostering better institutional links with the Arab highly skilled diaspora to better contribute to the transfer of knowledge, expertise, and skills to the sending countries. This may include as well establishing networks for the exchange of researchers and scientists in the field of migration between source and destination countries.

7. At The Research Level

There is a need to fill in the different gaps in our knowledge about the migration of Arab health professionals within and without MENA. There are many areas of suggested research that need to determined:

1. There is a need for new and improved data on population flows. In general, the data on migration in the region is outdated, fragmented, scattered, and in many cases unreliable.
2. There is a lack of data on remittances associated with the characteristics of migrants.
3. Regional organizations (including the Arab League, Arab Labor Organization, ESCWA) could play a significant role in setting a unified system for data collection, categorization, and definitions. This data could then be compiled and published as an annual labor market statistics book at the regional level.
4. Support training to researchers and professionals and develop a methodology to estimate and forecast migration flows , patterns, and general migrant characteristics.

Note

1 The ELMPS 06 sample contains 8,351 households. 3,685 of these households are original ELMS 98 survey households. 2,168 households split from original ELMS 98 households (for example, due to marrying and forming a separate household). In addition, 2,498 new households were added to the sample to form a refresher sample. 72% of individuals interviewed in 1998 were successfully re-interviewed in 2006, forming a panel that can be used for longitudinal analysis. ELMPS 06 contains information on international migration history as well as a module on current migrants and remittances. Include this in the text rather than in a footnote, per Harvard style standards.

References

Abu, EB (2014) Contribution of labour migration to development in sending and receiving countries. Paper presented to the Workshop on Arab Labour Migration and employment, Tunisia, 1-35

Akl, EA, Maroun, N., Major, S. et al. (2007) Graduates of Lebanese medical schools in the United States: An observational study of international migration of physicians. *BMC Health Services Research* 7: 7-49

Akl, EA, El-Asmar, K., Maroun, N. et al. (2008) Did the post war repatriation of Lebanese physicians drive recent Lebanese medical graduates to emigrate? An observational study, *BMC Health Services Research* 8: 1-9

Akl, EA, Maroun, N., Major, S. et al. (2008) Post-graduation migration intentions of students of Lebanese Medical schools: a survey study. *BMC Health Services Research* 8:1-8. Amna, A. and Mohamed, O. (2010) Highly-skilled migration: Sudan. Consortium for Applied Research on International *Migration (CARIM) Analytic and Synthetic Notes 2010/61*, Rome

Badr, ElS. and Abd ElRahman, A. (2006) Improving Health in Developing World: What can Medical Associations do? The Case of Professional Associations in Sudan. *Sudanese Journal of Public Health* 1(3):216-219

Badr ElS. (2011) Migration of Health Professionals in Sudan: Need for a National Policy? *Sudanese Journal of Public Health* 6 (3):1-2

Central Agency of Public Mobilization and Statistics (CAPMAS) (2014) *Bulletin of Permanent Migration.* Cairo: Central Agency of Public Mobilization and Statistics

Central Agency of Public Mobilization and Statistics (CAPMAS) (2013) *Bulletin of Permanent Migration 2000-2013*. Cairo: CAPMAS

Central Agency of Public Mobilization and Statistics (CAPMAS) (2013) *Labour Force Sample Survey.* Cairo: CAPMAS

Central Agency of Public Mobilization and Statistics (CAPMAS) (2000) *Labour Force Sample Survey.* Cairo: CAPMAS

Central Agency of Public Mobilization and Statistics (CAPMAS) (2012) *Quarterly Labour Force Sample Survey (LFSS): 2nd Quarter 2012.* Cairo: CAPMAS

Central Agency of Public Mobilization and Statistics (CAPMAS) (2007) *Statistical Database*. Available: www.capmas.gov.eg

Chaaban, J. and Mansour, W. (2012) The Impact of Remittances on Education in Jordan, Syria & Lebanon. ERF Working Paper (684): 1-34

Clemens, MA (2015) Losing Our Minds? New Research Directions on Skilled Migration and Development CGD Working Paper 415, Center for Global Development.

Clemens, MA and Pettersson, G. (2008) New Data on African Health Professionals Abroad. *Human Resources for Health*, 6:1-11

Di Bartolomeo, A., Fakhoury, T. and Perrin, D. (2010) CARIM – Migration Profile Lebanon. Available at: www.carim.org/public/migrationprofiles/MP_Lebanon_EN.pdf

Di Bartolomeo, A., Jaulin, T., Perrin, D. (2010) CARIM – Migration Profile Sudan, 2010. Available at: www.carim.org/public/migrationprofiles/MP_Sudan_EN.pdf

Economic & Social Commission For Western Asia (ESCWA) (2009) *Compendium of Social Statistics And Indicators International Migration Statistics in the ESCWA Region* Beirut: ESCWA

Egypt Demographic & Health Survey 2008

Elbadawy, A. and Roushdy, R. (2009) Impact of International Migration and Remittances on Child Schooling and Child Work: The Case of Egypt, Paper submitted to the ERF 16th Annual Conference on Shocks, Vulnerability and Therapy, Cairo.

El-Jardali, F. et al. (2008) Analysis of Health Professionals Migration: A Two-Country Case Study for the United Arab Emirates and Lebanon, Report Submitted to World Health organization

El-Jardali, F., Dumit, N., Jamal, D., et al. (2008) Towards a Better Understanding of the Magnitude and Predictors of Nurse Migration out of Lebanon, Report submitted to the Eastern Mediterranean Regional Office of the World Health Organization

ElShahrani, A. (2012) Population Composition and its impact on Security in GCC Countries, Riad: Gulf Cooperation Council (in Arabic)

Fouad, Y. ,Fahmy, Y., Abdel Hady, S. and Elsabagh, A. (2014) Egyptian future physicians are packing to leave but may be willing to return, *International Health* 7(3):190-194

Gulf Cooperation Council (2011) *Statistical Book* Riyad: Gulf Cooperation Council

Gulf Labour Markets and Migration Statistical Database. (2015) Available at: http://gulfmigration.eu/tag/statistics/

Haque Md. and Pitea R. (2013) Arab youth and labour mobility in International Organization for Migration (IOM), Arab Labour Organization (ALO) & Partners in Development (PID) *Intra-Regional Labour Mobility In The Arab World.* Cairo: IOM, 41-60

European Communities (2000) *Push and Pull Factors of International Migration: A Comparative Report.* Luxembourg: Office for Official Publications of the European Communities

ILO (2009) International Labour Migration & Employment in the Arab Region: Origins, Consequences & the Way Forward. Thematic Paper Presented to the Arab Employment Forum, Beirut.

Martiskova, M. (2013) What are the main reasons for young Greeks to emigrate? Master Thesis, University of Twente, School of Management and Governance

Moataz, Y., Khalifa Hussein, D., Samuilova, M. et al. (2012) *Egypt: Mobility of Health Professionals.* Brussels: International Organization for Migration

Mobility of Health Professionals (MOHPROF), National profile of migration of health professionals – Egypt, Available at: http://www.mohprof.eu/LIVE/national_profiles.html

Mourshed, M., Hediger, V. and Lambert, T. Gulf Cooperation Council Health Care: Challenges and Opportunities. Available at: http://tinyurl.com/hybagxg

Munzoul, A. (2010) Highly-skilled Sudanese migrants: gain or drain? Consortium for Applied Research on International *Migration* (CARIM) Analytic and Synthetic Notes 2010/13, Rome

Nair, M. and Webster, P. (2012) Health professionals' migration in emerging market economies: patterns, causes and possible solutions. *Journal of Public Health*, 35(1): 157-163

Nassar, H. (2006) Migration: A perspective from the MENA region, Paper presented to the seminar on «Strengthening Research Capacities to Enhance the Benefits of Migration for Development, Center of Future Studies, Stockholm

Nassar, H. and Ghoneim, A. (2003) Trade and migration: Are they complements or substitutes: A review of four MENA countries. *Center for Economic and Financial Research and Studies Economic Studies* Working paper 28/2003, Cairo

Naufal, G. and Genc, I. (2014) The story of remittance flows from the GCC countries. Gulf Labour Markets and Migration Explanatory Note (5/2014)

Organization of Economic and Cooperation Development (OECD) (2010) Database on Immigrants in OECD Countries (DIOC) Available at: http://tinyurl.com/z8hjp97

Packer, C., Labonté, R. and Runnels, V. (2009) Globalization and the cross border flow of health workers. In Labonté R. Schrecker T. Packer C. Runnels V. (Eds) *Globalization and Health: Pathways, Evidence and Policy.* U.K.: Routledge, 213-234

Tabar, P. and Rassi, R. (2013) The case of lebanon as an origin &destination country of arab labour mobility in International organization for migration (IOM), Arab Labour Organization (ALO) & Partners in Development (PID) *Intra-Regional Labour Mobility In The Arab World.* Cairo: IOM. 86-108

Policy Brief for the Global Policy Advisory Council (2010), *The Gulf Cooperation Council (GCC) and Health Worker Migration.* Available at: www.aspeninstitute.org

Shah, N. (2005) *Restrictive Labour Immigration Policies in the Oil Rich Gulf: Implications for Asian Countries.* Available at: http://iussp2005.princeton.edu/download.aspx?submissionId=50842

Stewart, J., Clark, D. and Clark, PF (2007) Migration and recruitment of healthcare professionals: Causes, consequences and policy responses, Hamburg Institute of International Economics Focus Migration Policy Brief (7), Hamburg

Suter, B. (2005) Labour Migration in the United Arab Emirates, Field Study on Regular and Irregular Migration in Dubai, Master Thesis, Malmö University, Sweden

Straubhaar, T. (2000) International mobility of the highly skilled: Brain gain, brain drain or brain exchange», Hamburg Institute of International Economics Discussion Paper (88), Hamburg World Bank (2015) Migration and remittances: Recent developments and outlook. Special Topic: Financing for development, *Migration and Development Brief* (24): 1-24

World Bank (2011) *Migration and Remittances Factbook 2011*. Washington D.C.: World Bank

World Bank (2010) *Middle East and North Africa Region Labor Migration from North Africa: Development Impact, Challenges, and Policy Options*. Washington: World Bank

World Bank (2008) *Economic Developments and Prospects: Middle East & North Africa Region; Regional Integration for Global Competitiveness* Washington D.C.: World Bank

World Health Organization (WHO) (2010) Migration of health workers, WHO Fact Sheet No.(301), Available at: http://www.who.int/mediacentre/factsheets/fs301/en/index.html

Zohry, A. (2005) *Interrelationships between Internal and International Migration in Egypt: A Pilot Study*, Submitted to Forced Migration & Refugee Studies Program, American University in Cairo

Zohry, A. (2006) *Attitudes of Egyptian youth Towards Migration to Europe*, Report Prepared for the Information Dissemination on Migration Project (IDOM), Cairo.

Appendix

Fig. 1: Migrant Remittance Inflows (US$ Million) to the Arab Region by Country (US$ Million) - 2014

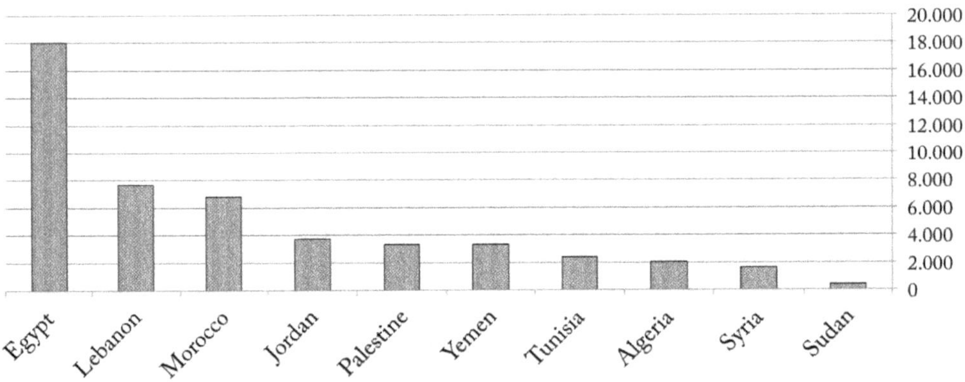

Source: World Bank 2015

Table 1: Expatriates in OECD by Country of Origin

Country of Birth	Emigrant Population	Education		Emigration Rate	
		Primary	Tertiary	Total	Tertiary Educated
	0	(%)	(%)	(%)	(%)
United Arab Emirates	14,4	22,0	25,4	0,6	...
Djibouti	5,4	34,6	30,2	1,3	...
Algeria	1.316,8	55,6	16,4	6,2	15,4
Egypt	313,7	19,8	48,3	0,7	3,7
Iraq	338,7	42,2	28,6	2,3	8,4
Bahrain	7,2	16,4	41,6	1,5	5,3
Jordan	63,9	20,3	41,5	2,1	4,6
Kuwait	37,1	17,0	45,2	2,2	6,5
Qatar	3,3	16,6	45,2	0,7	...
Saudi Arabia	34,1	23,4	36,8	0,3	...
Sudan	42,1	24,3	41,4	0,2	4,6
Lebanon	337,8	35,3	31,9	12,5	...
Libya	65,3	45,3	23,7	1,8	...
Palestine	15,5	25,5	44,0	0,9	...
Morocco	1.510,1	62,3	14,2	7,2	...
Somalia	125,1	50,5	14,4	3,1	...
Syria	130,2	34,1	34,5	1,3	3,8
Mauritania	15,2	63,5	17,3	1,0	...
Tunisia	430,3	55,9	16,1	6,1	14,3
Oman	2,6	14,2	39,2	0,2	...
Yemen	31,9	48,7	20,0	0,3	...

Source: Database on Immigrants in OECD Countries (DIOC), Barro & Lee 2000

Table 2-a: Net Migration Rate in GCC Countries (Net Number of Migrants for Each 1000 of the Population)

	Qatar	U.A.E	Kuwait	Bahrain	Oman	K.S.A
1950-55	42,6	0,1	19,4	0,3	-4,2	0
1955-60	25,7	0	32,3	1,1	-3,7	0
1960-65	62,9	65,8	67,4	7	-3,3	4,8
1965-70	68,3	60,9	46,7	-5,5	-1,4	7,2
1970-75	65	138,8	20,5	14,1	6	15,7
1975-80	37,7	102,4	26	20,4	14,3	20,6
1980-85	65,1	39,7	13,2	6,7	11,8	24,9
1985-90	29,5	29,3	18	7,9	1,1	14,4
1990-95	5,7	31,6	-61,8	9,2	2,2	-5,8
1995-2000	15,1	40,4	32,1	5,4	-6,1	0,7
2000-05	58,4	31,5	21,5	5,8	-4	2,6
2005-10	93,9	15,6	8,3	5,2	1,4	1,2

Source: El Shahrani 2012

Table 2-b: Total Population and Percentage of Nationals and Non-Nationals in GCC Countries (National Statistics, 2010-2013) (with Numbers)

Country	Date/ Period	Total Popu-lation	Nationals	Non-nationals	% in Total Population	
					Nationals	Non-Nationals
Bahrain	mid-2014	1.314.562	630.744	683.818	48	52
Kuwait	31-Mar-15	4.161.404	1.283.726	2.877.678	30,8	69,2
Oman	25-Mar-15	4.149.917	2.324.327	1.825.590	56	44
Qatar	April 2010	1.699.435	243.019	1.456.416	14,3	85,7
Saudi Arabia	mid-2014	30.770.375	20.702.536	10.067.839	67,3	32,7
United Arab Emirates	mid-2010	8.264.070	947.997	7.316.073	11,5	88,5
Total		50.359.763	26.132.349	24.227.414	51,9	48,1

Source: Gulf Labour Markets and Migration Statistical Database

Table 2-c: Distribution of Expatriates in GCC Countries by Country of Origin (2013)

Number				
	Arab Countries	Asian Countries	Developed Countries	Grand Total
Bahrain	128377	536171	5772	670320
Kuwait	356814	1490200	16043	1863057
Oman	61122	981013	7953	1050088
Qatar	281703	1176513	12666	1470882
Saudi Arabia	2567445	5986929	66896	8621270
United Arab Emirates	1393061	5817962	62632	7273655
Total GCC	4788522	15988788	171962	20949272
Percentage				
	Arab Countries	Asian Countries	Developed countries	Grand Total
Bahrain	19,20%	80,00%	0,90%	100,00%
Kuwait	19,20%	80,00%	0,90%	100,00%
Oman	5,80%	93,40%	0,80%	100,00%
Qatar	19,20%	80,00%	0,90%	100,00%
Saudi Arabia	29,80%	69,40%	0,80%	100,00%
United Arab Emirates	19,20%	80,00%	0,90%	100,00%
Total GCC	22,90%	76,30%	0,80%	100,00%

Source: Author's calculations based on United Nations, Department of Economic and Social Affairs 2013

Table 2-d: Distribution of Nurses in Ministry of Health (U.A.E) by Nationality (%) (2007)

Country	Number	%
India	1.097	31,4
UAE	839	24,0
Egypt	296	8,5
Philippines	276	7,9
Sudan	177	5,1
Jordan	142	4,1
Palestine	141	4,0
Oman	106	3,0
Somalia	82	2,3
Syria	66	1,9
Yemen	55	1,6
Lebanon	45	1,3
Pakistan	43	1,2
Tunisia	35	1,0
Indonesia	20	0,6
Iran	13	0,4
Iraq	12	0,3
Algeria	7	0,2
Others	40	1,1
Total	3.492	100,0

Source: El-Jardali et al. 2008

Table 3: Distribution of Health Workers in GCC Countries by Nationality (2009)

	U.A.E	Bahrain	K.S.A	Oman	Qatar	Kuwait	Total
Doctors in Public Hospitals							
Nationals	------	910	11.841	------	810	2.855	16.416
Non-Nationals	------	291	28.295	------	3.227	3.539	35.352
Doctors in Private Hospitals							
Nationals	------	------	828	------	------	------	828
Non-Nationals	------	------	15.939	------	------	------	15.939
Doctors in Governmental Clinics							
Nationals		289	1.441	------	------	------	1.730
Non-Nationals	------	84	5.412	------	------	------	5.496

Source: Gulf Cooperation Council 2011

Table 4-a: Distribution of Egyptians Residing Abroad by Country of Residence According to Latest Data Available from the Egyptian Ministry of Foreign Affairs, 2009

Country of Destination	Number	(%)
Arab Countries	4.789.359	74,00
African Countries	2445	0,04
Asian Countries	6073	0,10
Australia	106.000	1,60
European Countries	790.799	12,20
Americas	780.841	12,10
Total	6.475.517	100,00

Source: Ministry of Manpower 2009

Table 4-b: Working Permits for Egyptians by Country Groups (2001-2013)

Total	Foreign Ships	European Countries	African Countries	Asian Countries	Arab Countries	Country Group Year
574909	3325	17556	701	311	552709	2001
100	0,58	3,05	12,00	0,05	96,14	%
611414	3384	19608	799	299	587023	2002
100	0,55	3,21	13,00	0,50	96,01	%
648624	3629	24546	755	323	619042	2003
100	0,56	3,78	12,00	0,50	95,44	%
734232	3922	28984	963	486	699517	2004
100	0,53	3,95	13,00	0,70	95,27	%
784912	4652	29675	912	480	748849	2005
100	0,59	3,78	12,00	0,60	95,41	%
876423	5635	31032	1009	466	837780	2006
100	0,64	3,54	12,00	0,50	95,59	%
1008772	6841	33728	913	511	966234	2007
100	0,68	3,34	9,00	0,05	95,78	%
1100033	5979	34123	1133	570	105658	2008
100	0,54	3,1	11,00	0,05	96,15	%
1118474	5482	33943	1310	728	1076494	2009
100	0,49	3,03	0,12	0,06	96,25	%
1131199	5602	32440	1425	706	1090575	2010
100	0,5	2,87	0,13	0,60	96,41	%
1295623	4493	30020	721	352	1259779	2013
100,00	0,35	2,32	0,06	0,03	97,23	%

Source: CAPMAS various years

Table 5-a: Distribution of Work Permits Issued for Egyptians Working Abroad by Education and Country Group - 2013

Total	Below Intermediate & Without Qualification	Intermediate	Above Intermediate	University and Above	
1259779	353927	547383	58979	299490	Arab Countries
30020	6819	16179	1324	5698	European Countries
13	2	1	-	10	Australia
721	50	141	15	515	African Countries (Non-Arab)
352	18	45	9	280	Asian Countries (Non-Arab)
245	29	50	3	163	Americas
4493	411	1865	272	1945	Foreign Ships
1295623	361256	565664	60602	308101	Total

Source: CAPMAS 2013

Fig. 2: Distribution of Work Permits Issued for Egyptians Working Abroad by Education and Country Group - 2013

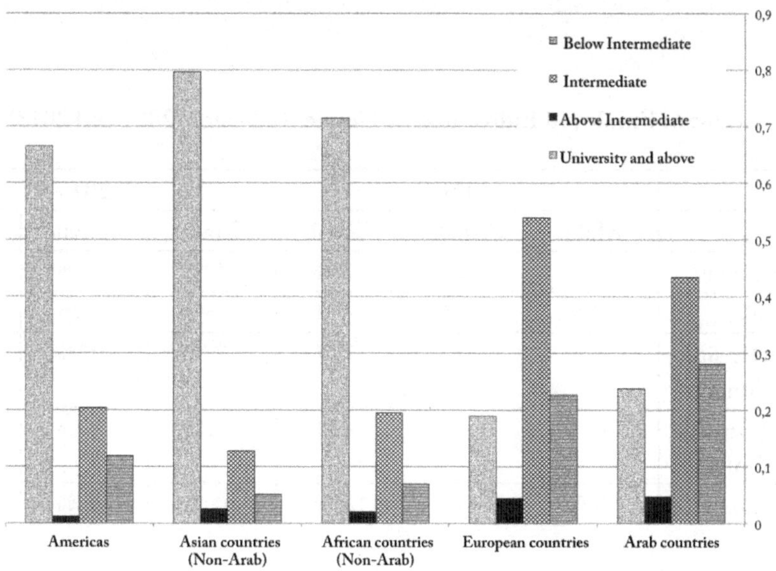

Source: Table 5-b

Table 5-b: Distribution of Permanent Migrants by Level of Education (2000-2013)

	P.H.D	Masters	High Diploma	University	Below Intermediate, Intermediate & Above Intermediate	No Qualification	Total
2000	8	12	8	243	233	32	563
2001	5	14	12	322	334	31	718
2002	10	10	11	274	323	24	652
2003	4	2	2	167	104	14	293
2004	3	16	9	200	115	14	357
2005	5	15	9	196	180	20	425
2006	4	12	7	210	124	11	368
2007	3	10	10	198	145	7	373
2011	5	9	11	300	224	29	578
2013	2	8	6	231	138	24	409
Total	49	108	85	2341	1920	206	4736
Share (%) (2000-2013)	1,0	2,3	1,8	49,4	40,5	4,3	100

Source: CAPMAS Database

Table 6: Employed Individuals by Gender & Sector (2001 and 2013)

	2001			2013		
	Male	Female	Total	Male	Female	Total
Government	25%	42%	28%	18%	37%	22%
Public	7%	3%	6%	5%	2%	4%
Private (Inside Establishments)	26%	20%	25%	27%	13%	24%
Private (Outside Establishments)	42%	34%	42%	47%	47%	47%
Total	100%	100%	100%	100%	100%	100%

Source: Author's Calculations based on CAPMAS-B various issues

Table 7-a: Distribution of Egyptian Migrants Who Have Acquired Immigrant Status by Occupation (When Leaving the Country) and Country (2013) (Number & Percentage)

%	Total	Others	Italy	Australia	Canada	U.S.A	
4,7	19	-	1	2	8	8	Legislature and Mangers
41,1	166	2	4	16	95	49	Specialists
2,5	10	-	2	1	2	5	Specialists Assistants
2,0	8	-	2	-	1	5	Clerical
1,2	5	-	2	-	-	3	Sales and Services
1,0	4	-	2	-	-	2	Artisans
1,0	4	-	3	-	1	-	Production Workers
46,5	188	8	98	7	39	36	Unclassified
100,0	404	10	114	26	146	108	Total

Source: CAMPAS 2014

Table 7-b: Distribution of Egyptian Migrants Who Have Acquired Immigrant Status by Scientific Qualifications (When Leaving the Country) and Country (2013) (Number & Percentage)

%	Total	Others	Italy	Australia	Canada	U.S.A	
5,7	14	1	-	-	11	2	Medicine
0,8	2	-	-	-	2	-	Dental
15,4	38	1	-	9	24	4	Pharmacy
2,4	6	-	-	-	5	1	Veterinary medicine
0,8	2	-	2	-	-	-	Agriculture
4,0	10	-	-	1	7	2	Science
22,3	55	1	-	6	27	21	Engineering
28,7	71	2	3	8	30	28	Commerce
3,2	8	1	2	-	1	4	Law
3,2	8	-	3	-	2	3	Arts
1,2	3	-	1	-	1	1	Education
12,1	30	2	8	-	7	13	Other High Level Qualifications
100,0	247	8	19	24	117	79	Total

Source: CAMPAS 2014

Table 8: Physicians & Nurses Born in Egypt Appearing in Census of Selected Receiving Countries circa 2000

Destination Country	Physicians	Nurses
Domestic	143.555	187.017
U.K.	1465	108
U.S.A	3830	661
France	471	89
Canada	750	45
Australia	535	87
Total Abroad	7.119	992
Fraction	5%	1%

Source: Clemens, Michael & Pettersson 2008

Table 9-a: Professional Distribution of Highly-Skilled Sudanese Migrants

Profession Category	Number of Migrants in Each Category
University Professors	2406
Medical Professions	10253
Engineers	12373
Administrative Professions	86471
Legal Professions	
Labourers	451443
Teachers	4360
Unspecified Contracts	175179
Total	744606

Source: Secretariat for Sudanese Working Abroad 2009 cited in Assal 2010

Table 9-b: Physicians & Nurses Born in Sudan Appearing in Census of Selected Receiving Countries circa 2000

Destination Country	Physicians	Nurses
Domestic	4.973	26.730
U.K.	606	42
U.S.A	65	85
France	17	12
Canada	15	20
Australia	40	7
Total Abroad	758	166
Fraction	13%	1%

Source: Clemens, Michael A. and Pettersson 2008

Table 10: Estimates and Projections for Remittance Flows to Developing Countries ($ Billions)

	2009	2010	2011	2012e	2013f	2014f	2015f
All Developing Countries	316	341	381	406	438	482	534
East Asia and Pacific	85	95	106	114	123	136	152
Europe and Central Asia	37	37	41	41	45	51	58
Latin America and Caribbean	57	58	62	64	68	75	84
Middle-East and North Africa	34	41	43	47	50	53	56
Share in Total Remittances to Developing Countries (%)	10,8	13,0	13,6	14,9	15,8	16,8	17,7
South Asia	75	82	97	109	118	130	144
Sub-Saharan Africa	28	29	31	31	33	36	39

Source: World Bank 2012

Table 11: Migrant Remittance Inflows (US$ Million) to the Arab Region by Country ((US$ Million))

Migrant Remittance Inflows (US$ Million)	2009	2010	2011	2012	2013	2014e	Remittances as a Share of GDP in 2013 (%)
Algeria	2.059	2.044	1.942	1.942	2.000	2.063	1,0
Egypt, Arab Rep.	7.150	12.453	14.324	19.236	17.833	18.000	6,6
Jordan	3.465	3.517	3.368	3.490	3.643	3.754	10,8
Lebanon	7.558	6.914	6.913	6.918	7.551	7.674	17,0
Morocco	6.269	6.423	7.256	6.508	6.882	6.819	6,6
Sudan	1.394	1.100	442	401	424	429	0,6
Syrian Arab Rep.	1.350	1.623	1.623	1.623	1.623	1.664	...
Tunisia	1.964	2.063	2.004	2.266	2.291	2.362	4,9
West Bank and Gaza	1.198	1.509	1.666	2.060	2.520	3.344	...
Yemen, Rep.	1.160	1.526	1.404	3.351	3.343	3.343	9,3

Source: World Bank 2011

13

'Outsiders' in France, 'Westerners' in the Gulf: Motives for Expatriation in the Professional Trajectory of Second-Generation French Graduates of North African Descent

Martin Lestra and Elyamine Settoul[1]

Introduction

Three French nationals settle in Dubai every day (Arabia Talent, 2015) while 2,000 French citizens reside in Qatar. In 2,000, around 11,000 French arrived in Bahrain; in 2012 there were nearly 37,000.[2] This article analyzes the contemporary migratory phenomenon of a subgroup of French nationals - French graduates of North African descent - to the Gulf states. Our contribution lies at the intersection between two developments in sociology and migration studies.

First, discrimination against 'visible minorities' (Moreau, 2010) has acquired visibility in the field of social sciences. Central both to researchers and policy-makers, the professional integration of French nations of North African descent is under scrutiny. Discrimination against the latter has reached structural levels: it has become relatively constant (Jones, 2013). Discrimination translates into limited access – as result of widespread discrimination practices – to their integration in the French labour market (Beauchemin et al, 2010). Indeed, it is estimated by some studies that being of North African descent can decrease by 40% the chances of being recruited in the French labour market (Adida et al., 2010; Edo and Jacquemet, 2014). Against these negative experiences of hiring discrimination and the phenomenon of the glass ceiling (Federal Glass Ceiling Commission, 1995), increasingly, young graduates decide to take their chances abroad. When many traditionally went to Europe or North America, one observes the emergence of new flows towards the South.

Undeniably – and secondly, studies of labour migrations are uncovering the rise of mobility around the globe – and, particularly, the increasing relevance of South-South and North-South migrations (Beaugé, 1985; Facchini et al., 2013). Flows of high-skilled workers and ensuing 'networks' of transnational knowledge – the "brain gain" (International Organisation for Migration, 2005) are considered crucial by all countries in want of further economic development (EUobserver, 2015).[3] The Gulf region embodies these developments. It is structurally in need of imported labour (see Figure 1).

Figure 1: GCC: Percentage of Nationals and Non-Nationals in Employed Population in some GCC Countries (national statistics, latest year or period available)[4]

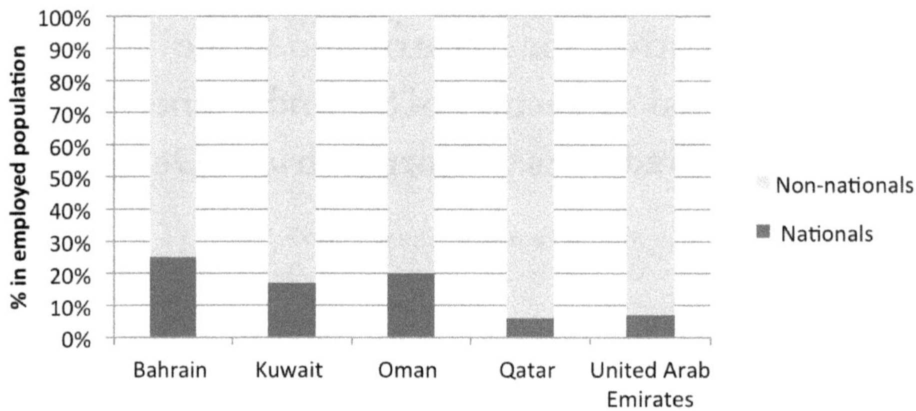

Non-nationals thus occupy working positions at all levels and across all sectors in these countries, from the business sector to the least explored sector, defense (GLMM, 2008; United Nations, 2013).

At the cross-roads of these research agendas, this article focuses on the mobility of French graduates of Maghreb descent in the Gulf states. Our chapter explores the underlying causes of this phenomenon with two questions. Why do French nationals of Maghreb descent desire to become expatriates in the Gulf states? How do these individuals – being less likely than others to be employed in France – manage to pursue successful careers in the Gulf?

The analysis of empirical data suggests that young French of North African descent benefit from the democratization and globalization of diplomas, competencies and labour markets. Secondly this population perceives the Gulf states as enabling them to occupy a favourable status (that of the "Westerner") attached to them. Crucially, these individuals manage to develop a career by making use of their spatial capital - "the sum of resources an actor has at its disposal which enable it to take advantage of society's spatial dimension depending on the strategy they choose" to and from the Gulf labour market (Kaufman, 2005).

Methodology and Scope of the Study

Made difficult by structural constraints – notably the legal provisions hindering the collection of ethnic-based data in France and the lack of disaggregated data published by national agencies in the Gulf states on the other, this research builds on a varied body of sources (Fargues, 2006). This implies that the chosen approach is on the one hand inductive. We draw on the experiences of individuals to identify motivating factors for their departure already present in the literature – and thus elaborate on the notion of spatial capital. On the other hand, it is also deductive as we confront expected explanations drawn from the literature[5] with expatriates' collected accounts.

The core body of our empirical investigation consists first of 25 semi-structured interviews of French nationals of Maghreb descent with professional experience in the Gulf region.[6] We also made use of all relevant primary administrative sources developed by the French state in parallel to the expansion of France's parliamentary representation to its expatriates communities around the world in 2012 (French Ministry of Foreign Affairs and Development, 2013).[7] The body of statistical and legal material collected in March-April 2015 from the Gulf Labour Markets and Migration database (Kuwait University and European University Institute), also provided us with a broader view on the migration and integration patterns of non-nationals in the Gulf region.

To enrich the available data we have also attended seminars and information sessions targeting students and professionals willing to emigrate to the Gulf. We have interviewed the directors of the two main non-state and non-profit providers of information and services on French expatriation to the Gulf: *Arabia Talent* (based in France) and *Association Hégire* (based in Dubai). They have provided us with complementary information on the trends observed in the professional trajectories of those who have used their services.

Table 1: Main Providers of Information and Counselling for French Expatriation to the Gulf Region (2015)

	Arabia Talent	Association Hégire
Members on Social Networks (Facebook, Linkedin, etc.)	4200	2410
Date of Creation	2013	2009
Mission	Counselling, information and job posting with respect to the Gulf for any individual wishing to work in the Gulf.	Counselling, information, networking events, sportive and cultural manifestations for the French-speaking Muslim expatriates in the Gulf.
Contact	http://www.arabia-talent.com/	https://www.linkedin.com/groups/Hegire-3381142/about

Source: compiled by authors[8]

With the exception of Saudi Arabia, we have focussed on the states of the Gulf Cooperation Council[9]. Indeed, individuals eager to work in that country having not been observed during this research.[10] Furthermore, Saudi Arabia has a very different history of migration than the other states of the GCC as it has not been under direct British rule (Kamrava and Babar, 2012). The choice to limit the study to individuals with high educational capital became self-evident during our empirical investigation. Attendance in informational meetings brought to the fore a self-selected professional profile. Theoretically open to all – these meetings consisted of participants between the ages of 25 to 34 years, a majority of whom were unmarried and held advanced curricular degrees.

Literature Review

Our contribution lies at the intersection of sociology and migration studies. First, discrimination against 'visible minorities' has acquired visibility in the field of the social sciences. Central both to researchers and policy-makers, the case of the professional integration of French of Maghreb descent is under scrutiny (Taguieff, 1987; Santelli, 2001; Silberman and Fournier, 2006; Madoui, 2008). Despite notable examples of successful professional achievements on the part of numerous French nationals of North African descent (Roy, 2015), discrimination against the latter has reached structural levels: it has become relatively constant (Frickey et al, 2005); and has now striking features. According to Ene Jones E. (2012), holding an advanced degree cannot compensate for the 'Maghreb stigma' on the job market. Such studies however do not engage with the impact of discriminatory practices on labour migration in the context of a globalized world. Rather, such insights are essentially journalistic (Rodier, 2014; Geoffroy, 2014) or normative (Marquardt, Achour and Achour, 2015) – in the case of France. When focused on French nationals of North African descent, studies are oft-centred on these individuals as Muslims. These contributions often point to the nature of their putative religious affiliations with Gulf centres of religious consciousness (Adraoui, 2013; Kazerouni, 2014). If religion and culture are part of the equation in this contribution, they do not preclude other explanatory factors.

Second and in parallel to this pattern of quasi-structural discrimination, mobility is on the rise in many regions of the world. Numerous studies exist on migrant flows in the 'South' and the wider MENA region is no exception to the trend (Hourani, 2010). However, studies of migrations to the Gulf focus – for notable numerical reasons – focus on low-skilled labour-intensive migrations (i.e. South-East Asian workers to the Gulf States) (Ali, 2011; Seshan, 2012; Gardner et al., 2013).

Only in rare instances does the literature address Southern competition for high-skilled workers and Gulf states' recruitment strategies (Kazerouni, 2014; Malecki and Ewers, 2007; Jendrissek, 2014; Davidson, 2014). Other studies focus on attitudes GCC citizens may have towards these migrants (Diop et al., 2012). Finally, non-nationals' integration in Gulf societies and professional organizations is dealt with in ethnographic accounts (Ali, 2010; Coles and Walsh, 2010), small case studies (Bhuian et al, 2001) or macro-level comprehensive assessments (Fargues, 2011). Less is known about the motivating factors for expatriation to the Gulf.

In this respect, we aim to contribute in three respects to the existing literature. First, this chapter refines our understanding of why high-skilled expatriates are attracted to the Gulf states. Second, our chapter assesses the implications of discriminatory practices on labour markets in the context of international labor flows. Third, this chapter contributes to the debate on the relevance of the North-South divide in the migration literature.

The Context of International Mobility: Background Information on the Gulf States and French 'Visible Minorities'

For the past fifteen years, the number of French expatriates has risen from a mere million in 2001 to 1 610 000 in 2014 – thus increasing by 3 or 4% every year in the past decade) (French Ministry of Foreign Affairs and Development, 2013). Out of these expatriates, 3% (about 48 300 people) reside in the Near East or Middle Eastern region. As a result of France's endeavours to internationalize its universities, most French degree programs require that students study or work abroad – this often constitutes the first experience with expatriation.

In parallel, the Gulf states' structural need for a qualified workforce compels them to seek it abroad. If the widespread presence of migrant labourers[11] is evident in the Gulf, the presence of "imported" qualified personnel is also characteristic of that region. GCC states recruit their highly-qualified personnel – by far – amongst Westerners (Fargues, 2011). Indeed, human resources needs are high at both ends of the qualification spectrum, and the nationalization of workforce policies instigated by Gulf authorities hides the fact that to this day, there is "little competition" between the nationals and foreign workers both for low- and high-skilled positions (Diop et al., 2012). In the UAE for instance, 15% of foreign migrants hold at least a university degree (Fromherz, 2012).[12] One priority of Qatar's development strategy laid down in its Vision 2030 is "the size and quality of the expatriate labour force" and the selected path of development" (Ministry of Development Planning and Statistics, 2008).

If the Gulf states' structural need for labour is crystal-clear, the organization of their labour market is complex. Professional stratification in the Gulf ties in with ethnic segmentation. Strikingly, labour stratification is part of the Arabic language as Gulf nationals distinguish the *moqeemen* (high-skilled) from the *ommal* (low-skilled) migrant workers (Diop et al, 2012). This static view does not reflect increasing in-group heterogeneity. Labour migrants from India are for instance increasingly highly-skilled while simultaneously providing Gulf states with its largest cohort of migrant labourers (Seshan, 2012);[13] Lebanese nationals have "tended to spread themselves fairly evenly in different sectors" (Hourani, 2010). Nevertheless, regional origin plays an essential role in defining the conditions of stay in the Gulf (Bel-Air, 2014). Hence, nationality is a determining factor of the atomization of the Gulf workforce along "ethnic and income-based lines" (Luomi, 2014). In that respect, French nationals are part of the Western-topped pyramid of migration labour – Asian nationals being at its bottom. Such stratification has implications for numerous aspects of daily life in the Gulf, be it wages, (Khalaf et al., 2015) living conditions, legal status, type and place of residence (Kamrava, 2012).

Considering the socioeconomic characteristics of the demographic demands of the Gulf states, it is not surprising that our research subjects were highly-qualified. If at the European level, Britons dominate the migration landscape in the Gulf, French nationals are – as in the case of Qatar – at times the second most common European cohort working in the region. Gulf states – notably Qatar - are also increasingly maintaining privileged relations with France and its 'visible minorities'. Ties between France and the Gulf states have been upgraded as well as diversified. Partnerships in the economic (arms sales), academic (the new campus of the Sorbonne University in Abu Dhabi), cultural (the Louvres Museum opened in Abu Dhabi) (Kerr, 2015) and sportive (Qatar's investment in the Paris Saint-Germain football club) sectors are flourishing.

In parallel to state-level relations, one can also observe occasional and tentative attempts to develop transnational ties between Gulf states and visible minorities in France. It appears that the October and November 2005 riots in the French 'banlieues' (Ossman, 2006; Goaziou and Mucchielli, 2007; Kepel, 2012) (suburbs, in particular surrounding the capital city of Paris) have acquired large visibility abroad. During this period, specific countries, including Qatar, but also the United States and Sweden, have shown interest in developing partnerships with entrepreneurs, social workers or promising leaders in the Parisian suburbs (Bernard, 2013). For instance, Qatar was strikingly visible in offering to fund a delegation of the ANELD (National Association of the Elected Representatives of the Diversity) with 50 million euros worth of entrepreneurial projects within the French 'banlieues'.

In short, the structural needs of the Gulf labour market coincides with the increased globalization of graduates' university curricula on the one hand; and the tightening of links between the petro-monarchies and the French state and minorities on the other. Thus, having provided context to the professional expatriation of French nationals of North African descent, we now explore the empirical evidence for their motivations.

Empirical Findings – Interviews

Interviewees' Profiles: Mobile Graduates with Spatial Capital

Our interviewees share a double characteristic: they are all children who benefited fromthe democratization of access to education and the internationalization of higher education. In other words, they derive from a relatively humble background - most of their parents were blue collar workers who arrived in France during the migration surges of the 1960's. They took advantage of the 1990s' opening up of universities to a larger public (Beaud, 2003). All of them hold a Bachelor's or a Master's degree in trade, the industrial sector, hotel management and catering, or the communication sector (Arabia Talent, 2015). Second, interviews show that they have a tendency to be geographically mobile. Through a student exchange or a professional experience abroad, all of them have had an international experience prior to their departure to the Gulf region (for relatively long periods of time, from 6 months to a year). Three-quarters of the interviewees have benefited from the internationalization of their university curriculum. In particular, they took part in the Erasmus Program and dwelt in other European countries. Thus Youssef explains:

> I left for a whole year to the University of Amsterdam and then stayed there for three years. I improved my English and I saw a different society.

Confrontation with another ethno-cultural diversity and to a cosmopolitan spirit also emerges as a strong memory in the interviews:

> I stayed for two years and a half near London. There I enjoyed people's mentality in the workplace. People don't look at how you live or how you're dressed. Money is money. I worked as a contractor for the development of an ICT projects for big firms like IBM and Sony. I also worked for Quantum. My manager was a Sikh who wore a turban. At the canteen you could see veiled Muslims and Jews eating at the same table. Nobody cared. There was even a prayer room for those who wanted. Ask for the same thing in France? They'll look at you weirdly at work. (Mourad)

For that matter, they possess a "spatial capital" understood as "the sum of resources an actor has at its disposal which enable it to take advantage of society's spatial dimension depending on the strategy they choose" (Kaufman, 2005). They are 'globalization's offspring' in the sense that they do not hesitate to make use of one or the other side of their Arabo-Western identity to maximize their integration and opportunities.

In his appraisal of cultural openness, Mourad also points implicitly to another aspect of the interviewees' profile: they have become more and more mobile. This aptitude comes out strongly in Karim's account of his Erasmus experience in the United Kingdom:

> With the low-cost companies everything changed. I became a pro of the low cost. As an earlybird, I found tickets to go to the UK for less than 20 euros. It became cheaper for me to fly to Conventry than to go back home by train to my parents' place in Thionville in Eastern France.

In an analogy to Bourdieu's theoretical framework, a few researchers have offered the concept of "mobility capital" to describe this competence. Lévy and Lussault (2003: 124) thus define it as "the array of resources, accumulated by an actor, and enabling him to take profit on, depending on his strategy, of the spatial dimension of society".

Therefore, our interviewees are the offspring of the democratization and internationalization of higher education. Our interviewees have made strategic use of these experiences to develop their spatial capital, to acquire skills (including but not limited to linguistic ones) and have come to appreciate the Anglo-Saxon societies that they consider to be more progressive and tolerant. These characteristics thus tie in closely with the motives that have led these individuals to depart to the Gulf.

Why Leave from France to the Gulf Region?

In this section we contrast existing explanatory frameworks in the migration literature, with the subjects' own perceptions. Does professional expatriation to the Gulf result from a desire for professional fulfilment and the wealth associated with this region of the world (Appadurai, 2001)? Does discrimination and growing islamophobia in France also matter (Geisser, 2003; Delthombe, 2005)? Is there – beyond economic calculus – a desire to draw nearer to the Arabo-Muslim culture (Piscatori and Dresch, 2005)?

Interviews show that four main factors have influenced their choices to leave for the Gulf from the most widely cited to the least cited.

1. The Weight of Dreams: the Gulf Success Story

First and foremost, the Gulf region is synonymous with personal and professional accomplishment for all interviewees. The region is known for its economic boom. As recalled by Appadurai, "fantasized migrations" are a substantial condition preceding migration itself. As collective representations, they nourish individuals' initiatives. For others like Gildas Simon, "migration fantasies are enshrined in the numerous myths and stories of the Odyssey, the big discoveries or the Eldorado (…) that are opposed to a daily routine perceived as unbearable (…) feeds into the cultural common good of humanity" (Colin, 2008). For Abelès:

> Imagination invades daily practices, notably in situations of migration, in which subjects are required to invent an exile into their own world, making use of all the images put forward in particular by the media. (2008: 38)

Forged by comic books, films and advertisement, the stereotypical images of Gulf states' wealth are a strong inciting factor. Leaving for these countries symbolizes – automatically – the promise of economic wealth. More importantly, the economic boom experienced by these countries is put into contrast with the stark economic situation in France and Europe at large. Thus Karim explains:

> Things now happen in the South. In France or in Europe, you can say whatever you like, but it's only 'humming'. Each one maintains his own privileges, there is

> nothing else to do. Here [in the Gulf] everything is to be done, there are projects and budgets, and that is what I find interesting.

Other interviewees display a geopolitical understanding of Gulf states' evolution:

> These countries want to weigh in on things. They are in to soft power and they want to show the world that they can organize the biggest events like Formula 1 or the World Cup. Today, to go to Asia you nearly always need to stop in Dubaï. Same goes for Islamic finance and universities. They want to be at the top and they have the means to get there. That's why the challenge is so interesting. It's exciting.

Though the region's dynamic reputation serves as a major incentive, it is not the only mechanism at work.

2. Going Beyond the Glass Ceiling: Turning the Arab-Muslim "Stigma" into a Professional Advantage

If fantasy is a motivating factor for departing to the Gulf, the weight of multidimensional discrimination also sustains that decision. Many leave for the Gulf hoping to find recruitment processes that apply the principle of equality of opportunity. Some have recounted stories of discriminations they underwent in France as well as of the glass ceiling that prevented them from achieving the status or position they felt that they merited. In a basic definition, Pierre-André Taguieff recalls that racial discrimination designates the exclusion of individuals from the profit of shared social goods due to their race, or a group-belonging perceived to be of a racial type (Taguieff, 1987). One identifies this phenomenon when the principle of equality of treatment is undermined in an illegal manner. Neglected for long, racial discrimination has since the work of Philippe Bataille in the 1990s come under closer scrutiny within French sociological circles. Bataille described the direct and indirect methods through which discrimination can affect professional careers, both at their initial and advanced stages (Bataille, 1997). Research on the professional integration of young French graduates of Maghreb descent has revealed the existence of widespread discrimination (Beauchemin et al, 2010). Some studies go further by demonstrating that the possession of a higher diploma does not suffice to "compensate" for the negative connotation of the "North African" stigma (Jones, 2013). Despite being oft-mentioned in French public debates, our interviewees present these daily encumbrances as commonplace. First, their names and ethnic descent are clearly mentioned as preventing them from climbing up the social and professional ladder. Inès, a project manager for the Qatar Foundation thus states:

> I have worked in luxury hotels in Paris as well as in Southern France. I was competent and motivated in fulfilling my tasks, I speak four languages but I was soon told that I was going to "hit the ceiling" and that I would never obtain the managing position I was aspiring to. Roughly speaking, a Maghrebine could not manage an entire team for a major hotel company because this does not correspond to the company's habits nor to the image it wishes to project of France towards foreigners. I don't regret having left for Doha. Today I have responsibilities and a wage that I could never have had in France. Here everything must be done, from one week to the other we can be solicited or hired by other companies. Al Jazeera has recently approached me, I've also set up my own structure in parallel to my job.

If Inès describes a glass ceiling preventing her from exploiting her competence and full potential within a group that does not wish to promote a French national of North African descent outwards, Tewfik, having worked in the banking sector, underlines, beyond racial discriminations, the "codified" nature of French recruiting systems:

> In reality in France everything is codified. We like people who come out from the "grandes écoles" but not from universities. To be selected for an interview, it is better to be the worst from a "grande école" than the best from a good university. It's a double-speed system. If you add to this racism against those bearing an Arab name and the lack of professional networks, you can understand that we find ourselves falling by the wayside.

Lofty, who works in the import-export sector in Bahrein subscribes to these views when describing French recruitment systems:

> In France we look at which school you're from rather than what you're really capable of doing. That's why I prefer the Anglo-Saxon mentality. When working in Australia, the only thing that mattered was 'how much do you bring to the company'? There, we were rather evaluated on concrete results. In France, they'll look at your school and then base your wage grid on that. Actually, we put a leash on those who are 'hungry'.

The degree issue was also often cited. In France, degrees are hierarchically ordered within each sector and according to the interviewees, recruiters emphasize diplomas rather than experience. Some of the interviewees believe that France restrains their competencies and talents.

Aware of their qualifications, they hope to be able to fully exploit their potential in Gulf countries. Some consider the latter to be closer to Anglo-Saxon countries, which are considered to be less discriminating than France in hiring procedures and professional meritocracy. Finally, notwithstanding their professional experiences, all interviewees consider affirmative action policies in France to be insufficient. Thus Farid testifies:

> Who do you want to complain to? That's the unnerving hypocrisy in France, everyone knows that Arabs and Blacks are put aside of managing positions but there is no way to resist. I've got a friend who bluntly decided to change the name on his business card. He's called Hassen but he asks to be called Arsène. We've got to accept this reality or go away. I've decided to leave.

Lotfi confirms:

> Most of my buddies have left to London, Canada or Luxembourg because they know that they'll be left aside for promotions back in France. I had a friend who appealed to the High Authority Against Discrimination and for Equality (HALDE) for racism while recruiting. He was the best of all candidates and was "squeezed out" by recruiters. What did the HALDE do about it? I don't know, nothing much I imagine. Since then he's created his own business and has a lot of success.

In addition to the fact that agencies designed to fight discrimination lack credibility or "bite", one first notes that graduates attempt to find individual solutions to their professional difficulties (expatriation, entrepreneurship, name change, etc.). All of these mechanisms take place as if in conformity with the "republican" ideal, their mind-set has interiorized the inability to organize themselves as a community to claim their rights. *Consequently, all answers to their problems are upheld by individual endeavours rather than collective ones.* Second, there appears an overlap between the existence of widespread discrimination in the recruiting process and the perception at large that professional opportunities are limited in France – at least compared to those offered in the Gulf.

Indeed, subjects' perceived sense of discrimination translates into their wish to make their Arab identity a professional advantage rather than a stigma. From their perspective, the Gulf offers them that opportunity. Indeed, being Arab and Muslim might actually be an advantage in these countries because of shared cultural characteristics and religious beliefs. Such views are held by expatriates, such as Idris, who after failing multiple interviews in France, is now working in Dubai:

> When I first applied for companies in France, I felt that I had the experience but that the company was reluctant because they see you as something new. They don't want to take risks because they've never recruited people from the

> Maghreb. In Dubai, the whole world is working, you meet Indians, Chinese, Arabs. It's globalization in an Arab country. I'd applied on Bayt.com, a website specialized on the Gulf region. So we all have our chance and as a Muslim we even have an advantage because we know Islam, its values, the Ramadan, etc.

Most of the interviewees are Muslims that respect religious rites (Ramadan, not drinking alcohol) and some do their daily prayers. For them, living in a mainly Muslim society is considered to be easier than maintaining a professional and family life in the West. In a way, they make use of their dual identity to show their adaptation skills. What is considered to be a disadvantage and a stigma in the French professional context is turned into an advantage in countries where cultural and religious references are Islamic.

The advantages evoked above echo these expatriates' knowledge and internalisation of Muslim normative codes, entailing diverse behaviours, from gender relations to festivities or again, Ramadan. Our interlocutors have the feeling that what appeared to as a negative maker – or what Goffman defines as a "stigma" – transforms into a resource taking into account their religious and cultural environment. At the least, their cultural and religious specificity tends to become invisible in their social practices in the Gulf. Thus, Lofti draws on the comparison between his professional experiences in France and in the Gulf region as regards fasting for religious motives:

> In France when you're following Ramadan, all of your colleagues ask you questions. Why do you do this? Isn't it too difficult? What's the purpose? Every year you have to explain it all over again, the principle, that it's a pillar of Islam, and all of this is tiring over the longer term.

Others, like Farid, evoke such practice as stemming the tide of their professional careers:

> I was a salesman for many years in France. When you invite a client for lunch and explain that you don't drink alcohol or that you don't eat because of Ramadan, it creates immediately a relational problem. If you don't eat while he's eating and while you're negotiating, then the relationship is not the same and, regardless, it creates a barrier between us two and has an impact on your sales revenue.

Others go further to argue that one has to dissimulate such practices at work, in her case, in the business district of La Défense, in Paris:

> I had Muslim friends at La Défense who never said that they were following Ramadan and made arrangements to pray "in secret" at their workplace. That's

> the way it is, companies are used to it and so as not to create trouble we remain discrete. But there again I feel sorry to have to almost hide because of this.

These excerpts point to several phenomenon. Lofti concentrates on French ignorance of Islamic practices, while Farid describes alcohol and foods as pathways of sociability. For him, alcohol consumption creates social bounds and contributes to the fluidity of interpersonal relationships. Similarly, adhering to the fasting requirements of Ramadan, appears as an obstacle to professional performance. On the contrary, such religious prescriptions mesh with societies of the Gulf, who live in accordance to the rhythm of Muslim traditions. This is what Inès explains:

> Here in Qatar we live with the pace of Muslim countries and so everything is simpler professionally-speaking. We follow Ramadan, and companies adapt their working hours. We go out of work earlier for example. There are prayer rooms in every professional sector. It's clear that everything is simpler and this is valuable.

3. Avoiding the Impact of the Socio-Political Climate

The third reason is not the determining factor but is a recurrent element of this empirical study. Interviewees systematically mentioned their perception of an Islamophobic climate gaining momentum in France and Europe at large in the last decade. This is presented in two ways that are not mutually exclusive.

The first concerns the repeated, disproportionately present and negatively biased debates on Islam or Political Islam. Mourad (bachelor in Islamic Studies) underlines this phenomenon and describes himself as "tired" of the multiple debates about Islam in the last decade:

> We've have had… I don't know how many debates in France. On national identity, on the burqa, on the minaret, on hallal… Muslims are always targeted in these debates. And at some point, you say stop. We authorize homosexual marriage but forbid covering our hair. It's too much.

For his part, Fouad criticizes the broader debate on national identity introduced by Eric Besson in 2009, former Minister for Immigration and National Identity. According to Fouad, this national conversation represents a disguised question on Muslim integration.

> The government pretended it was launching a real reflection on national identity in France but we all know who this really targeted. The real problem for Muslims is

> not the burqa, terrorism but the possibility to work and be treated like the others. Because without a job you don't exist and you fall back onto your own identity.

The issue of professional integration and the triviality of islamophobia in the French political arena is widely shared. Vincent Geisser describes the various forms that islamophobia adopts. Their common denominator is the consideration of religion as an obstacle to the process of social and cultural assimilation implicitly valued by social actors. In his analysis, Geisser (2003) describes the multiple reign of islamophobia à la française. The discourse pertains to a wide spectrum of actors going from 'hardcore' anticlerical republicans ("secular fundamentalists") to innumerable political movements close to the extreme-right).

The second concern expressed is the media "treatment" of Islam-related events. Tariq insists on the particular newsworthy presentation of Muslim populations:

> Sometimes in France you watch the news on BFM TV, and you'll see three or four news pieces on Arabs, violence, terrorism and the "banlieues".[14] When you see that every day, for months, it starts to rot your mind.

Nabil, a technician in the media sector in Dubaï is outspoken on this issue:

> After 2005, television only spoke of Islam in the 'banlieues'. After the riots, we had debates on national identity, the minarets, the veil, hallal, and the lonely wolves. All that ties to the Muslim religion could be subject to debate. That kind of atmosphere disgusted me. I don't pay taxes to be spat on.

Our interviewees unanimously condemn the negative streak of public and newsworthy debates on Islam, the focus on the *niqab*, given the very limited scope of such a practice (Mandraud, 2009). They criticize the "double discourse" making a minority's minor practice a national priority and reversely, dissimulating widespread and commonplace discrimination practices. They resent the Muslim 'stigma' that they consider associated with them in the media. Along these lines Assia points out:

> When I learn that a terrorist attack has taken place in a Muslim country, I'm always afraid that an Arab or a Muslim is responsible for it, because I know that this will bounce back to me when it comes to reputation (…). There will be discussions at work, remarks on Muslims. It comes back incessantly and I'm almost obliged to apologize to be of the same origin as these people who do that kind of bullshit.

Tying international geopolitics to French Muslim individuals is a frequent issue raised by the interviewees. For them, this phenomenon is accentuated by the role played by the media who seek to increase their audience, using spectacular stories of which news items tied, from a distance or close up, to Islam, are part (Bourdieu, 1996).

4. Reconnecting with their Cultural Origins

Being held as a plausible hypothesis, the desire to reconnect with one's cultural origins does not constitute a dominant motive in these work migrations. A small number of interviewees explicitly see their professional mobility in the GCC countries as an opportunity to reconnect with their Arab or Muslim culture of origin. Choosing a country of Arab or Muslim culture is an opportunity to bring them closer to their origins. Most of the interviewees have limited knowledge of Arabic – mostly spoken dialects from Maghreb (Algerian, Moroccan and Tunisian) – and cannot read or write it.[15] Most of them are only able to speak a dialect from Maghreb. Some of them, such as Lofti and Mourad, were able to take Arabic classes during their childhood in mosques or French schools that had partnerships with consulates of Maghreb countries.[16] However, the teaching quality varied greatly and was often disappointing. Farida tells us the following:

> Our Arabic is remote from the true Arabic. Most people in the Middle East don't understand the spoken [dialect] from the Maghreb because people in the streets have modified it so much. Moroccans or Algerians are far away from their own language. The worst is Algerian because it has many French words that have been mixed with the language.

Beyond the limited Arabic teaching classes on offer, it is the relationship of France with the Arabic culture and language that is questioned. The schooling system is thus put into question. Myriam:

> I've got many friends who want to learn Arabic and they can't because there is nothing on offer. So they turn towards classes offered by mosques and private institutes.

Holding a bachelor's degree in Islamic studies and a student of Arabic, Mourad laments the lack of consideration given to this language in the French public schooling system:

> The Arabic language is not valued enough. It's strange because many young people have an interest in it. It's often easier to learn Russian or Chinese than Arabic while it's spoken in many countries on the shoreline and with direct contact with Europe.

Several reports and articles confirm this statement with namely two elements. The first is the unsuited content of Arabic teachings provided through the ELCO programme. Trained for the most part in Maghreb, the teachers are not well integrated within the broader teaching staff. Moreover, in secondary schools Arabic professors are mostly left to the position of replacement teacher, or requested to teach in various institutions, because of the limited budget allocated to Arabic. In addition to these administrative problems is the oft-encountered hostility of head teachers. The latter view offering Arabic classes as risky because it may attract schoolchildren of North African descent perceived to be "difficult". In other terms, enrolling individuals of Arab descent would be synonymous with an overall degraded reputation for these educational institutions (Dichy and Reymond, 2014) as Arab populations are often associated with socially challenging environments, and "ghettos". *Sales manager* in Dubai, Abdallah shares this point of view and wishes to give his offspring linguistic opportunities:

> I want to give to my children the schooling opportunities I've never had. I will put them into an international school and have them learn French, English and Arabic at a very young age. French for home, English for work, and Arabic for the religious culture.

Discussion

The Power of Gulf Narratives

These empirical findings provide evidence to support the powerful image the Gulf has for a specific audience. Interviews very clearly demonstrate the role played by imagination in their choice of destination. The economic pull of Gulf countries continues to operate and many interviewees have remotely looked for information on the Internet (Bayt.com, Dubbizle, on top of the aforementioned services of information) in order to grasp how these countries work as well as the differences among them. If the main way to go abroad used to be through a change of position within a firm, new ways of canvassing the market have appeared. As indicated by Fouad Kemache (Director of Arabia Talent), these graduates go so far as travelling to these countries on a tourist visa in order to directly get in touch with local companies. Thus, whether or not a Gulf strategy exists to recruit certain groups of individuals is not the matter at stake here. Gulf states do resonate positively with the interviewees. For instance, Dubai provides the Gulf ideal-type described by Kazerouni (2007: 21) as "a model of economic success seemingly disconnected from hydrocarbons and built on the employment of young talents from the Arab world." Hence, more than wages or financial leverage, Gulf states are made attractive because they offer the possibility to those working there of "contribut[ing] to the region's development, shap[ing] other socio-political imaginaries, and provide a basis from which they can claim to be in and of the Gulf' (Kamrava and Babar, 2012: 35).

The Paradox of Nationality: Shifting from "Outsider" to "Westerner" Status [17]

We have seen on the one hand that nationality is a dominant criterion for a professional's integration into the Gulf labour market. Previous studies had indeed shown that in the Gulf, sharing an "Arab" ethnicity is not necessarily a privilege, and has even historically been associeated with negative perceptions from nationals (Dorsey, 2014: 1740). The paradox here is enlightening. While French nationals of North African descent wish to turn an 'Arab stigma' into a professional resource, their French passport, as well as the French diploma – as confirmed by Fouad Kemache - enable them to benefit from the status of "Westerners" in the GCC.

Conclusion

This article, based on an unusual body of interviews and primary data, has investigated the motives for the professional expatriation of French nationals of North African descent. It built on existing scholarship on the sociology of 'visible minorities' and migration studies to build a transnational account of this subgroup's mobility.

Firstly, the available data underlines the crucial role of migratory imagination. It also stresses subjects' wishes to transform a stigma into a professional advantage. These pull factors are reinforced by a push factor: the perception of strengthened religious intolerance in France and in Europe. Finally, the will to reconnect with cultural origins is mentioned by some subjects. Thus, the Gulf states appear as states where these individuals can enhance their prestige. First, these migrants enjoy a favourable status (that of the Westerner), which dictates the advantageous material conditions of their stay in the Gulf. Second, these individuals consider that they are caable of taking advantage of the Arab stigma and transforming it into a resource within the host country's labour market. As observed, despite some entrepreneurs' efforts to pool efforts and information independently of French authorities, these individuals' strategies remain individually oriented. In other words, limited professional integration in France of specific subgroups may result in the transfer of resources to other economies – here the Gulf region. Secondly, it outlines the internationalization dynamics at work on the labour market, as well as in the realm of the market for competencies and diplomas. The young French of North African descent we have met fully epitomize this globalizing trend and the development of professional mobility from the North to the South. In that respect the notion of "North-South" does not make sense as the South is a heterogeneous reality encompassing diversities both in human and economic development. Among choice destinations of the South, Gulf states are privileged as a result of their economic attractiveness.

Notes

1 The authors share first authorship for this article. The order of names is made alphabetically.

2 *Note*. From *Gulf Labour Markets and Migration Programme, January 2013 (GLMM, 2013)*

3 *Note*. This is not only true of the Gulf region. In the European Union, efforts to make the EU attractive to high-skilled individuals lead to preferential visa access under the EU Blue card procedure. This trend is likely to be reinforced in the foreseeable future, as attractiveness of high-skilled workers adds to structural demographic deficits in various regions of the world. These elements are discussed in Honor Mahony, "EU to Overhaul 'Blue Card' Work Permit For Migrants," *EUobserver*, Brussels. May 13, 2015. Accessed September 5, 2015. https://euobserver.com/justice/128703

4 *Note*. Information collected by the GLMM programme was done so at different dates: Bahrain (April 2010), Kuwait (December 2013), Oman (December 2012), Qatar (2013), United Arab Emirates (May 2009).

5 *Note*. In particular our questionnaires reflect the following claims: that expatriation of French nationals of North African descent is due to a professional fulfilment and the wealth associated with this region of the world; to discrimination and islamophobia in France; or the will to draw nearer to the Arabo-Muslim culture.

6 *Note*. Semi-structured interviews lasted from 30 minutes to an hour-and-a-half. Questionnaires were systematically elaborated though questions were added or retrieved according to subjects'reactivity. Other studies on migration groups in the Gulf draw on larger samples. Seshan (2012) for instance, gathers 128 examples; the Qatar University in its Omnibus Survey 2012, has collected interviews from 851 resident expatriates (SESRI, 2012: 13). Our universe of cases is more restricted given the specific group. In Boolean terms it consists of: [French AND North African descent AND Professional experience in the Gulf]. Within this subgroup, interviewees were chosen to represent the diversity of gender, age, professional and family status, and second country of origin (of North Africa). Given the length of interviews, the limited size of the overall number of cases, as well as the overarching views shared by the founders of the most visible consultant firms on expatriation to the Gulf for the targeted populations, we find it likely that the material we have been working with is comprehensive in capturing the views this group. Second, because this article wishes to extend past research with new insights on the dynamics of migration and the resources used by migrants to reach their professional goals, we find it unnecessary to expand the number of interviews or to define control groups.

7 *Note*. Simultaneously to this administrative reform, the French Ministry of Foreign Affairs has initiated in 2008 an annual survey on French expatriates – its results are exploited in this article.

8 *Note*. Based on interviews and online fact-checking.

9 *Note*. United Arab Emirates, Oman, Qatar, Kuwait, Bahrain. It appears clear at the macroeconomic level that French expatriation is particularly salient in the resource-rich emirates of the UAE and Qatar.

10 *Note*. Indeed, Saudi Arabia is not the preferred destination for French nationals of North African descent's expatriation. However, Saudi Arabia is increasingly valued for its religious environment. Thus, a few cases are observed of resettlement of expatriates in Saudi Arabia - after a first experience in Dubai or Qatar. Hence, future studies of Western migration to the Gulf are likely to give a more prominent role to the Kingdom.

11 *Note*. In this article the terms of "migrant labourer" (or ommal) is used. Despite it being a misleading denomination that ascribes an ethnic dimension to a professional low-skilled task, it is however dominant in the literature; in addition, the authors do not intend to debate such a notion; hence its multiple use in this work.

12 *Note*. Numerous are the examples of high-profile Gulf institutions whose senior management is substantially composed of non-nationals: let's mention the Abu Dhabi Investment Agency is "staffed with individuals hailing from more than 40 countries" (Fromherz, 2012), 92; and the Qatar Foundation's decision-making Strategic Planning Directorate, which was composed in 2012 for half by non-nationals (Qatar Foundation, 2013)

13 *Note.* Data confirm that in-group heterogeinity strongly varies between nationals of different regions. For instance, 88% of Europeans employed in the UAE in 2008 occupied higher professional positions (clerk, senior officials, managers, etc) while this was only the case for 23% of all Asian (non-Arab) migrants; the same is observed with Kuwait (2014): 60% for Europeans and 11% or non-Arab Indians. (proportions are less significative in Kuwait because the number of non-stated professional status is much higher than in the case of the UAE) (Seshan 2012; GLMM, 2014)

14 *Note.* 'Banlieues' literally designates the "suburbs" and has a wider meaning describing the socially difficult, economically poor and ethnically segmented geographical suburbs around the largest French cities, notably Paris, Lyon and Marseille.

15 *Note.* Some interviewees have joined the Hégire Association in upon arrival in Gulf countries in order to learn Arabic together.

16 *Note.* Following a European Directive (July 25, 1977), France had launched the ELCO programme (Teaching Languages and Civilizations of Origin). This was organized between France and countries of emigration, at the level of primary schools, with the aim of reconnecting migrants' children with the culture of their country of origin.

17 *Note.* The status of "Westerner" is the most referred to in our interviews. It may be connected with the notion of "whiteness" utilized by Neha Vora (2012) in her postcolonial analysis of expatriation and (non)-integration of high-skilled personnel in the Gulf states. Her analysis corroborates our own empirical findings in the sense that French national of North African descent are first and foremost considered according to their passport than their ethnicity ; just as Vora's Indian origin is counterbalanced and superseded by the "whiteness" attached to her American passport.

References

Oumma TV. (2014) "*Ces Français Musulmans Qui Réussissent Aux Emirats.*" Available from: http://www.agoravox.tv/tribune-libre/article/ces-francais-musulmans-qui-45675 [10 January 2017].

The National. (2015) *Former British Foreign Minister Says Gulf Countries Must Do More over Syria*. Available from: http://tinyurl.com/zqwbuvj [10 January 2017].

The National. (2015) *High Living Costs Force Half of Expats to Consider Leaving UAE.*" Available from: http://tinyurl.com/jd76sjb [10 January 2017].

Diop, A. & Tessler, M. & Trung Lec, K. &Al-Emadid, D. and Howelle, D. (2012), Attitudes Towards Migrant Workers in the GCC: Evidence from Qatar. *Journal of Arabian Studies,* vol. 2, pp. 173-187.

Adida, C., Lantin, D. and Valfort, M. (2010) Identifying barriers to Muslim integration in France. *Proceedings of the National Academy of Sciences,* vol. 107, pp. 1-7.

Adraoui M-A (2013) *Du Golfe aux banlieues: le salafisme mondialisé.* Paris: Presses Universitaires de France.

Ali, S. (2011) Going and Coming and Going Again: Second-Generation Migrants in Dubai. *Mobilities,* vol. 6, pp. 553-68.

Syed, Ali. (2010) In Dubai, the Party Is Far From Over". *The Guardian* 20 March. Available from: http://tinyurl.com/grq4eks [10 January 2017]

Anam, T. (2014) Migrant Worker Nations, Unite! The New York Times 4 April. Available from: http://tinyurl.com/gujos99 [10 January 2017].

Appadurai, A. (2001) *Globalization.* Durham, NC: Duke University Press Books.

Arabia Talent, (2015) *Le Guide de l'Emploi dans les Pays du Golfe.* Paris: Arabia Talent.

Bataille, P. (1997) *Le racisme au travail.* Paris: La Découverte, 1997.

Beauchemin, C., Hamel, C., Lesné, M. and, Simon, P. (2010) Les Discriminations: Une Question de Minorités Visibles. *Populations et Sociétés,* vol. 466 pp. 1-4.

Beaud, S. (2003) *80% au bac et après? Les enfants de la démocratisation scolaire.* Paris: La Découverte.

Beaugé, G. (1985) Des Migrations Nord-Nord Aux Migrations Sud-Sud. *Revue Tiers Monde,* vol. 26, pp. 485-491.

Bel Air, F. (2014) *Demography, Migration, and Labour Market in Qatar.* Available from: Gulf Labour Markets and Migration [/August 2014].

Bernard, C. (2013) Avant le Qatar, la Suède S'Intéresse Aux Banlieues Françaises," *Slate* 10 October. Available from: http://tinyurl.com/l2bxrbg [10 January 2017].

Bhuian, S., Al-Shammari, E. and Jefri, O. (2001) Work-Related Attitudes and Job Characteristics of Expatriates in Saudi Arabia. *Thunderbird International Business Review,* vol. 43, pp. 21–32.

Body-Gendrot, S. and Wihtol De Wenden, C. (2014) *Policing the Inner City in France, Britain, and the US.* Basingstoke: Palgrave.

Bourdieu, P. (1996) *Sur la television.* Paris: Liber-Raisons d'Agir.

Coles, A. and Walsh, K. (2010) From "Trucial State" to "Postcolonial" City? The Imaginative Geographies of British Expatriates in Dubai. *Journal of Ethnic and Migration Studies,* vol. 36, pp. 1317-1333.

Davidson, C. (2014) Expatriates and the Gulf Monarchies: Politics, Security And the Arab Spring. *Asian Affairs,* vol. 45, pp. 270-88.

Delthombe, T. (2005) *L'islam imaginaire: la construction médiatique de l'islamophobie en France, 1975-2005.* Paris: La Découverte.

Dichy, J. and Reymond, P-L (2014). "Il faut enseigner l'arabe dans le service public", *Le Monde* 2 November. Available from: http://tinyurl.com/jpm4dum [10 January 2017].

Dorsey, J. (2014) The 2022 World Cup: A Potential Monkey Wrench for Change. *The International Journal of the History of Sport,* vol. 31, pp. 1739-1754.

Edo, A. and Jacquemet, N. (2014) Discrimination d'Origine et de Genre à l'Embauche et Homéophilie Ethnique: Une Expérience de Testing Par Correspondance en France. *Etudes de l'INSEE - Economie et Statistique,* vols. 464-465-466, pp. 155-172.

Facchini G., Mayda, A. and Mendola, M. (2013) *South-South Migration and the Labour Market: Evidence From South Africa.* Centre for Economic Policy Research, vol. 9450.

Fargues, P. (2011) Immigration Without Inclusion: Non-Nationals in Nation-Building in the Gulf States. *Asian and Pacific Migration Journal,* vol. 20, pp. 273-292.

Fargues, P. (2006) International Migration in the Arab Region: Trends and Policies. *Proceedings the United Nations Expert Group Meeting on International Migration and Development in the Arab Region.* Available from: United Nations [18 May 2006].

Federal Glass Ceiling Commission (United States Department of Labour). "Solid Investments: Making Full Use of the Nation's Human Capital". Washington, D.C. (November 1995), 4.

Frickey, A., Primon, J., Borgogno, V. et al. (2005) *Jeunes diplômés issus de l'immigration, insertion professionnelle ou discriminations.* La Documentation Française.

Gardner, A., Pessoa, S., Diop, Al-Ghanim, K., and Harkness, L. (2013) A Portrait of Low-Income Migrants in Contemporary Qatar. *Journal of Arabian Studies,* vol. 33, pp. 1–17.

Geisser, V. (2003) *La nouvelle islamophobie.* Paris: La Découverte.

Geoffroy, R. (2014) Qui Sont les Expatriés Français?" *Le Monde,* 17 October. Available from: http://tinyurl.com/hsvmtw8 [10 January 2017].

Gildas, S. (2008) *La planète migratoire dans la mondialisation.* Paris: Armand Colin.

Goaziou, V. and Mucchielli, L. (2007) *Quand les banlieues brûlent...: Retour sur les émeutes de novembre 2005.* Paris: La Découverte.

Gulf Labour Markets and Migration programme (2015). *Demographic and Economic Module.* Available from: http://gulfmigration.eu/ [10 January 2017]

Hourani, Guita (2010) *Lebanese Migration to the Gulf (1950-2009).*" *Middle East Institute,* 19 April. Available from: http://tinyurl.com/jcgt9ku [10 January 2017].

International Organisation for Migration (2005) *Mainstreaming Migration into Development Agendas.* International Organisation for Migration, Geneva. Available from: http://www.ppl.nl/books/ebooks/IOM_Red_Book_08.pdf [10 January 2017]

Jendrissek, D. (2014) Brain Drain and Brain Gain: The Global Competition to Attract High-skilled Migrants / The Cultural Politics of Talent Migration in East Asia, *Journal of Ethnic and Migration*

Studies, vol. 40, pp. 2055-2057.

Jones, B. (2000) Multiculturalism and Citizenship: The Status of 'Visible Minorities' in Canada. *Canadian Ethnic Studies*, vol. 32, pp. 111-126.

Jones, E. (2013) Discrimination à l'Embauche des Jeunes en Île-de-France: un Diplôme Plus Elevé Compense-t-il une Origine Maghrébine? *Etudes de l'INSEE – Économie et Statistique*, vols. 464-465-466, pp. 173-188.

Kamrava, M. K. and Babar, Z. (eds) (2012), *Migrant Labor in the Persian Gulf.* New York: Columbia University Press.

Kaufman, V. (2005) Mobilités et réversibilités; vers des sociétés plus fluides?. *Cahiers Internationaux de sociologie*, vol. 118, pp; 119-135.

Kazerouni, A. (2014) Musées et Soft Power dans le Golfe Persique. *Pouvoirs*, vol. 152, pp. 87–97.

Kepel, G. (2012) *Banlieue de la république: Société, politique et religion à Clichy-sous-Bois et Montfermeil.* Paris: Editions Gallimard.

Kerr, S. (2015) "Abu Dhabi admits Louvre's Gulf outpost faces further delay; Arts projects. *Financial Times* 27 June 2015.

Khalaf, A., Al-Shehabi, O. and Hanieh, A. (2015) *Transit States: Labour, Migration and Citizenship in the Gulf.* London, Pluto Press.

Lévy, J. (2003) *Dictionnaire de la géographie et de l'espace des sociétés, Paris, Belin.*

Luomi, M. (2014) *The Gulf monarchies and climate change: Abu Dhabi and Qatar in an era of natural unsustainability.* Oxford: Oxford University Press.

Madoui, M. (2008) *Entrepreneurs issus de l'immigration maghrébine, de la stigmatisation à la quête de la reconnaissance sociale.* Paris: Mondes Contemporains.

Mahony, H. (2015) *EU to Overhaul 'Blue Card' Work Permit for Migrants*, *EUobserver* 13 May. Available from: https://euobserver.com/justice/128703 [10 January 2017].

Malecki, E. and Ewers, M. (2007) Labor Migration to World Cities: With a Research Agenda for the Arab Gulf. *Progress in Human Geography*, vol. 31, pp. 467–84.

Mandraud, I. (2009) La police estime marginal le port de la burqa", *Le Monde* 30 July.Available from: http://tinyurl.com/n2eq9b [10 January 2017].

Marquardt, Felix; Mokless and Achour, M. (2015) *Jeunes de France, votre salut est ailleurs: barrez-vous!,*" *Libération* 3 September.. Available from: http://tinyurl.com/9ljvjg6 [10 January 2017].

Ministry of Development and Planning and Statistics (2008) *Qatar National Vision 2030.* State of Qatar.

Ministry of Foreign Affairs and Development (2013) *Enquête sur l'Expatriation des Français*, Maison des Français de l'Etranger, France.

Moreau, P. (2010) 'Êtes-vous invisibles?," *Revue Argument*, vol. 1. Available from: http://tinyurl.com/jnda77o [10 January 2017].

Ossman, S. (2006) The French Riots: Questioning Spaces of Surveillance and Sovereignty. *International Migration*, vol. 44: pp. 5-22.

Piscatori, J. and Dresch, P. (2005) *Monarchies and Nations: Globalisation and Identity in the Arab States of the Gulf.* London; New York: I.B. Tauris.

Social And Economic Survey Research Institute(2012) *Annual Omnibus Survey: A Survey of Life in Qatar.* State of Qatar.

Rodier, A. (2014) Expatriation: rêves de jeunes diplômés," *Le Monde*. 2 June. Available from: http://tinyurl.com/k554adp [10 January 2017].

Roy, O. (2015) *La peur d'une communauté qui n'existe pas, Le Monde*, 9 January. Available from: http://tinyurl.com/zqll5yb [10 January 2017].

Santelli, E. (2001) *La Mobilité Sociale Dans l'Immigration: Itinéraires de Réussite des Enfants d'Origine Algérienne.* Toulouse: Presses du Mirail.

Scholz, C. (2010) L'Attractivité des Armées Auprès des Minorités Ethniques en Europe: Une Comparaison Allemagne, France, Grande-Bretagne, *Fiche de l'Irsem*, vol. 2, pp. 1-12.

Seshan, G. (2012) Migrants in Qatar: A Socio-Economic Profile. *Journal of Arabian Studies*, vol. 2, pp. 157–71.

Silberman, R. and Fournie, I. (2006) Les Secondes Générations Sur le Marché du Travail en France: Une Pénalité Ethnique Ancrée Dans le Temps," *Revue Française de sociologie*, vol. 47, pp. 243-292.

Simon, G. (2008) *La planète migratoire dans la mondialisation.* Paris: Armand Colin.

Taguieff P (1987) *La Force du Préjugé, Essai Sur le Racisme et Ses Doubles.* Paris: La Découverte.

United Nations Population Division. (2014) *World Population Series 2000-2013.* United Nations. Available from: http://esa.un.org/PopPolicy/PolicyIssues.aspx [10 January 2017].

United Nations. (2013) *World Migration in Figures: A Joint Contribution by UN-DESA and the OECD to the United Nations High-Level Dialogue on Migration and Development.* Available from: http://www.oecd.org/els/mig/World-Migration-in-Figures.pdf [10 January 2017].

Vora, N. (2012) Expat/Expert Camps: Redefining "Labour" Within Gulf Migration. In: Kamrava, M. and Babar, S. (eds.) *Migrant Labour in the Persian Gulf.* New York: Columbia University Press, pp. 170-197.

Wihtol de Wenden, C. (2001) "Un Essai de Typologie des Nouvelles Mobilités", *Hommes et Migrations*, vol. 1233, pp. 5-12.

About the Contributors

Ali A. Alraouf is an architect and urban planner interested in research and practice related to holistic sustainable built environments. Alraouf has held permanent and visiting teaching and research positions at regional and international universities. He published more than 80 journal articles, refereed papers, critical reviews, essays, in addition to books and book chapters. He serves on editorial boards and acts as a reviewer for international journals and conferences. He has delivered lectures and presentations in over 25 countries. He is the recipient of a number of awards and was selected as a member of the 2012 Excel campaign at Qatar University. He currently acts as Head of the Capacity building, Research and Development unit at QNMP at the Ministry of Urban Planning in Doha, Qatar.

Pia-Kristina Anderson is an anthropologist at the American University of Sharjah in the United Arab Emirates. She received her Ph.D. in Anthropology from the University of California at Berkeley, and has taught anthropology at both the American University in Cairo and at the University of California at Santa Cruz. Her research and teaching areas include culture contact, anthropological history, archaeology, cultural geography and the formation of identity. Dr. Anderson has extensive fieldwork experience in the Middle East, the Pacific Islands, and Nepal, amongst other regions.

Erik Baark is Professor at the Division of Social Science and the Division of Environment at the Hong Kong University of Science & Technology. He was awarded a Ph.D. from Lund University in 1986. He has published numerous books, including Lightning Wires: Telegraphs and China›s Technological Modernization 1860-1890 and articles in top international journals such as The China Quarterly, Research Policy, and the International Journal of Technology Management. His research interests are related to innovation systems and environmental policies in China and other East Asian countries.

Rasmus Gjedssø Bertelsen is a Danish international relations scholar. He is Professor and holds the Barents Chair in Politics at the University of Tromsø-The Arctic University of Norway. Rasmus grew up in Iceland and studied at the Universities of Copenhagen, Iceland, Geneva, Lausanne and Amsterdam. He holds a Ph.D. in international relations from the University of Cambridge during which time he also was a visiting Ph.D. candidate at Sciences Po Paris. Rasmus completed his first postdoc at the Harvard Kennedy School of Government on the soft power of American and French universities in the Middle East. This was followed by a Japan Society for the Promotion of Science postdoc at the Tokyo Institute of Technology and United Nations University-Institute of Advanced Studies where he expanded his research on Middle East higher education research to East Asia. He returned to Denmark to complete a postdoc at Aalborg University (Denmark) on Sino-Danish knowledge relations. His field of research is on the role of transnational knowledge relations in international politics, both present and past. More specifically, Rasmus conducts research on US-European-Middle East-East Asian university relations, Sino-Danish triple helix knowledge flows and Arctic science diplomacy.

Michael K. Corman is an adjunct professor in the Department of Sociology and Anthropology and Faculty of Nursing at the University of Prince Edward Island, Canada. Prior to this position, Dr. Corman held appointments as an assistant professor in Sociology and the UKCRC Centre of Excellence for Public Health Northern Ireland at Queen's University Belfast and the Faculty of Nursing at the University of Calgary in Qatar. His research and teaching interests include the sociology of health and illness, emergency medical services, institutional ethnography, aging, caregiving, autism spectrum disorders, health care work, sociology of the family, gender and sexuality, critical qualitative research strategies, and public health.

Gao Xudong is Associate Professor at the School of Economics and Management (SEM), Tsinghua University, and senior research fellow of the Tsinghua University Research Center for Technological Innovation. He also serves on the Telecom Economy Expert Committee of the Ministry of Industry and Information Technology, People's Republic of China. His major research interests are in technology strategy, management of technological innovation, and strategic management. He holds a Ph.D. in Management from the MIT Sloan School of Management, an M.A. in industrial economics from the Renmin University of China, and a B.E. in industrial management engineering from the Harbin Institute of Technology.

Daniel Goldman is an Advisor to Syntax LLC, an investment management firm, and Locus Analytics, a development-stage venture focused on creating a taxonomic cloud platform for large-scale portfolio management. He holds a B.A. from Columbia, an M.B.A. from Yale, and a J.D. from Georgetown, and serves on the Bretton Woods Committee.

Isam R. Hamza is Professor of Japanese Studies, "Marubeni" Chair in Humanities at Qatar University, State of Qatar. He received his Litt.D. in Japanese Studies from Osaka University, Japan in 1991. Prior to his current academic appointment, he was a Professor of Japanese Studies, head of the Japanese Language and Literature Department, and Vice-Dean of the Faculty of Arts, Cairo University, Egypt. He has also taught at Osaka University for foreign Studies, Osaka University of Education, Waseda University, and the Defense Academy of Japan. His publications have appeared in the Faculty of letters Bulletin of Osaka University, the Periodical of the Scientific Association for the History of Japanese Thought, Mitashi-gakkai at Keio University, Studies in Asian Cultures Periodical at International Christian University in Tokyo-Japan, and the Bulletin of the Faculty of Arts at Cairo University Egypt.

Jason Hickey contributed to this project while working as a Nursing Instructor at the University of Calgary in Qatar. He now works as an Assistant Professor of Nursing at the University of New Brunswick in Canada. Jason›s research interests include mental health promotion and nursing education, including inter-professional education.

Laura L. Matherly is an Associate Professor of Strategy in the College of Business at Zayed University, Abu Dhabi. She has a Ph.D. from The University of Tennessee. She has extensive executive and consulting experience in training and developing strategic plans to improve organizational effectiveness with companies such as Boeing Commercial Aircraft Group, Microsoft, and Weyerhaeuser Paper Company. She has taught at Pennsylvania State and Texas A&M Universities. Her current research efforts include achieving performance excellence in organizations through measurement and more recently, the study of nationalization and how to make workforce quotas successful in GCC countries.

Martin Lestra is a Ph.D. candidate at the Department of Social and Political Sciences, European University Institute (Florence). His dissertation deals with Gulf states› influence in international organizations in the realm of foreign aid. He was a visiting fellow at New York University – Abu Dhabi and Qatar University in 2015-2016.

Heba Nassar is a Professor of Economics at the Faculty of Economics and Political Science at Cairo University where she also directs the Center for Economic and Financial Research Studies. She has also taught at the Cairo Demographic Center and the Department of Economics at AUC. She has acted as a consultant for WHO, UNFPA, UNESCWA, UNICEF, the Ministries of Health in Egypt, Kuwait and Bahrain, the Center for Human Services, USAID, Petrich Associates and the Netherlands Cooperation Programme. Her research interests include human resource development and economic integration. She is also a policy affiliate at the Economic Research Forum and a research professor affiliate of the Social Research Center of the American University in Cairo.

Camilla Nellemann is owner of Kokan, which assists clients in analyzing Japanese affairs and communicating with Japanese partners. She is also a researcher at the Copenhagen Business School. Dr Nellemann holds a Ph.D. in International Business from Rikkyo University in Tokyo, Japan, on Toshiba's strategic response to the Fukushima nuclear crisis, for which she was awarded a MEXT scholarship. Before joining academia, she worked with investment banking and export promotion in Denmark and Russia. She also served as a linguist officer in the Danish Armed Forces and has been deployed in Kyrgyzstan and Afghanistan.

Neema Noori is an Associate Professor of sociology at the University of West Georgia in the United States. He received his Ph.D. from Columbia University in 2006. Prior to his current academic appointment, he taught at Columbia University and the American University of Sharjah in the United Arab Emirates. His publications have appeared in PS: Political Science and Politics, International Journal of Culture, Politics, and Society; Political and Military Sociology: An Annual Review, and Central Asia Survey.

Ebrahim Radhi is a productivity and innovation professional holding the positions of Principle Consultant at Jafcon for Productivity Improvement and Executive Director at Green Innova. He received his M.Phil. from the University of Cambridge in 2010. Specializing in business development, process re-engineering, and innovation commercialization, he has provided consultancy services for more than 120 organizations in Bahrain, UK and Dubai. He is a member of the Institute of Management Services and the Chartered Management Institute in the UK, and he previously worked for The Welding Institute, Laing O'Rourke and Vestas Holding.

Jean-Marc Rickli is the head of global risk and resilience at the Geneva Centre for Security Policy (GCSP) in Geneva, Switzerland. He is also a research fellow at King's College London; a non-resident fellow in modern warfare and security at TRENDS Research and Advisory in Abu Dhabi; an advisor at Gulf State Analytic and a senior advisor for the AI (Artificial Intelligence) Initiative at the Future Society at Harvard Kennedy School. Prior to these appointments, Dr. Rickli was an assistant professor at the Department of Defence Studies of King's College London and at the Joaan Bin Jassim Joint Command and Staff College in Doha. He was also an assistant professor at the Institute for International and Civil Security at Khalifa University in Abu Dhabi. Dr. Rickli received his PhD and MPhil in International Relations from Oxford University, UK, where he was also a Berrow scholar at Lincoln College. He co-edited *The Small Gulf State: Foreign and Security Policies before and after the Arab Spring*, Routledge 2016, "The Political Rationale and Implications of the United Arab Emirates' Military Involvement in Libya", in Dag Henriksen and Ann Karin Larssen (2016) *Political Rationale and International Consequences of the War in Libya*. Oxford University Press, pp. 134-154 and "New Alliances Dynamics and their Impact on Small GCC States." *Third World Thematic* (2016).

Elyamine Settoul is currently postdoctoral fellow at IRSEM, French Ministry of Defense. He has been a Jean Monnet Fellow at the European University Institute in Florence. He has a PhD in political science from the Institut d'Études Politiques de Paris: his thesis deals with the enlistment of military personnel of immigrant origin. His field of interest includes migration issues, ethnicity, discriminations and political Islam. His academic articles and reports include "You're in the French army now: institutionalizing Islam in the Republic›s army", Religion, State and Society, vol 43, 2015 and "Analyser l'immigration postcoloniale en milieu militaire : retour sur les enseignements d'une enquête ethnographique", Les Champs de Mars, Juin 2015.

Sheikha Shamma bint Sultan bin Khalifa Al Nahyan is passionate about the sustainable development of the United Arab Emirates and its people, sharing the vision of her great-grandfather, the UAE's founder, His Highness Sheikh Zayed bin Sultan Al Nahyan. She is committed to supporting her community, with a focus on initiatives that empower women and promote healing through reading. An avid researcher, Sheikha Shamma has delivered papers at academic conferences both locally and internationally in the fields of gender studies, knowledge transfer, consumer behaviour, Emiratisation, and financial literacy. She holds a Bachelor of Science in Business Administration from Zayed University and has successfully completed a Post Graduate Certificate in Sustainable Business at the University of Cambridge.